T0185273

Engineering the Knee Meniscus

Synthesis Lectures on Tissue Engineering

Editor
Kyriacos A. Athanasiou, *Rice University*

Engineering the Knee Meniscus
Kyriacos A. Athanasiou and Johannah Sanchez-Adams
2009

Engineering the Knee Meniscus

Kyriacos A. Athanasiou and Johannah Sanchez-Adams

ISBN: 978-3-031-01448-2   paperback
ISBN: 978-3-031-02576-1   ebook

DOI: 10.1007/978-3-031-02576-1

A Publication in the Springer series

SYNTHESIS LECTURES ON TISSUE ENGINEERING

Lecture #1
Series Editor: Kyriacos A. Athanasiou, Rice University

Series ISSN
Synthesis Lectures on Tissue Engineering
ISSN pending.

Engineering the Knee Meniscus

Kyriacos A. Athanasiou and Johannah Sanchez-Adams
Rice University

SYNTHESIS LECTURES ON TISSUE ENGINEERING #1

ABSTRACT

The knee meniscus was once thought to be a vestigial tissue, but is now known to be instrumental in imparting stability, shock absorption, load transmission, and stress distribution within the knee joint. Unfortunately, most damage to the meniscus cannot be effectively healed by the body. Meniscus tissue engineering offers a possible solution to this problem by striving to create replacement tissue that may be implanted into a defect site. With a strong focus on structure-function relationships, this book details the essential anatomical, biochemical, and mechanical aspects of this versatile tissue and reviews current meniscus tissue engineering strategies and repair techniques. We have written this text such that undergraduate students, graduate students, and researchers will find it useful as a first foray into tissue engineering, a cohesive study of the meniscus, or a reference for meniscus engineering specifications.

KEYWORDS

meniscus, tissue engineering, meniscus biomechanics, meniscus tears, functionality index, bioreactors, meniscus cells, meniscus vasculature, alternate cell sources

*To my mother, Anastasia, and my late father, Aristos,
for always inspiring me, trusting me,
and showing me the way to excellence. –KA²*

*To Mom, Dad, and Josh for all of your love and support.
To past, present, and future inquisitive minds for
motivating this work. –Johannah*

Contents

Preface

The inspiration for writing this monograph came from our desire to highlight important aspects of the knee meniscus as they relate to tissue engineering and outline historical and current advances in the field. Our goal was to create a cohesive, logical, and graphic work that guides the reader through a study of the meniscus through the eyes of a tissue engineer. We hope that undergraduate students, graduate students, and researchers alike will find this information useful as an introductory guide to engineering the meniscus or a reference for future engineering attempts. The first half of the monograph focuses on native characteristics of the knee meniscus including its biochemical makeup, mechanical properties, and the limited ability of the tissue to self-repair. The second half reviews tissue engineering attempts aimed at recreating meniscus biochemistry and mechanics to repair damaged tissue, and highlights important considerations for meniscus tissue engineers. Throughout the chapters particular emphasis is placed on the functional aspect of the meniscus and its importance in a tissue engineered construct. We hope that the overall outline of this book allows the reader to understand the complexity of the knee meniscus and the need for engineered meniscus tissue, while providing enough background to evaluate recent engineering attempts. By documenting the field of meniscus engineering in this way, we aim to educate and inspire innovation from current and future researchers.

Kyriacos A. Athanasiou and Johannah Sanchez-Adams
Rice University
March 2009

CHAPTER 1

Structure-Function Relationships of the Knee Meniscus

1.1 ANATOMY AND DEVELOPMENT

1.1.1 ANATOMY OF THE KNEE MENISCUS

Integrating vasculature, cells, and extracellular matrix molecules, the knee meniscus comprises two semicircular, wedge-shaped pieces fixed in place via a network of ligaments. The meniscus is a glossy white fibrocartilaginous tissue that is an important component in the normal joint, as shown in Figure 1.1. The human knee contains a medial meniscus and a lateral meniscus which are located

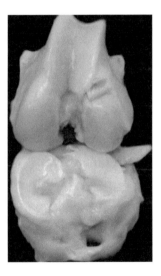

Figure 1.1: **The native knee meniscus.** Bovine knee joint showing the white, semicircular cartilages that make up the knee meniscus. The meniscus increases congruence between the femoral condyle and tibial plateau, and aids in normal joint function.

between the femoral condyle and tibial plateau (see Figure 1.2). The main functions of the meniscus are to increase congruency of shape between the curved condyle and flat plateau, maintain stability, and bear and transfer load within the joint. Because it functions in a joint articulation it exhibits

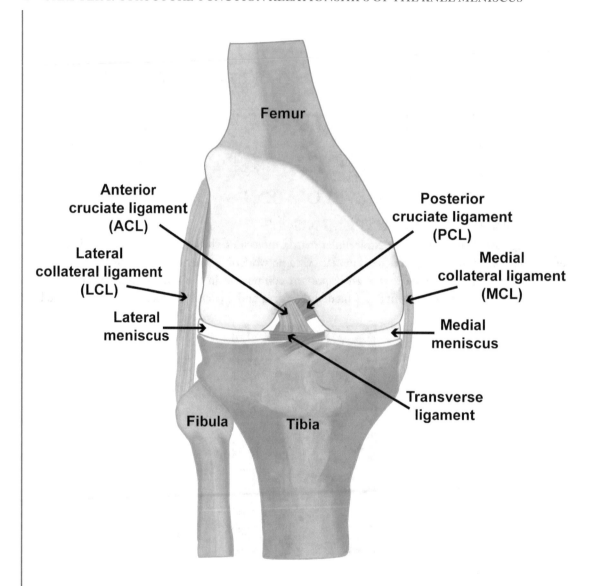

Figure 1.2: **Anterior view of the knee joint and location of the meniscus.** The meniscus is located between the femoral condyles and tibial plateau within the knee joint. It is made up of two parts, medial and lateral, which are attached to each other by the transverse ligament. Various ligaments within the joint space and on its periphery help to restrict bone movement, and maintain normal joint functionality. These include the posterior cruciate ligament, anterior cruciate ligament, lateral collateral ligament, and medial collateral ligament.

a smooth surface macroscopically and microscopically [1]. Both the medial and lateral menisci are wedge-shaped and semilunar, but the medial meniscus is generally more circular in shape than the lateral meniscus.

There is an array of ligaments that aid in stabilizing the meniscus within the knee joint during loading conditions, as shown in Figure 1.3. The ligaments of Humphrey and Wrisberg connect the posterior horn of the lateral meniscus to a lateral insertion site on the medial femoral condyle. The Humphrey's and Wrisberg ligaments are located anteriorly and posteriorly to the posterior cruciate ligament (PCL), respectively. Studies on human cadaveric knees have shown that an estimated 50% of people have both of these ligaments, while 93% have either one or the other. The medial meniscus is connected on its periphery to the medial collateral ligament which connects the femoral condyle to the tibia [2].

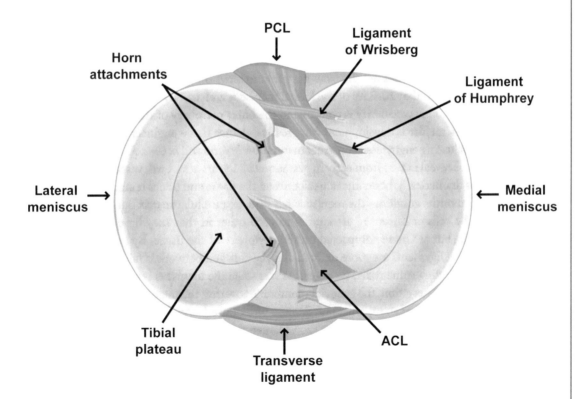

Figure 1.3: **Superior view of the tibial plateau showing meniscal attachments.** The medial and lateral menisci rest atop the tibial plateau and are affixed to the tibia via horn attachments and to each other via the transverse ligament. Other ligaments in the joint space help to restrict movement such as the ACL, PCL, and ligaments of Wrisberg and Humphrey.

The anterior portions of the meniscus are joined together by the transverse ligament, and each meniscus is anchored to the tibial plateau via anterior and posterior meniscal horns. The insertion sites of these horns are highly innervated and display four different zones that connect the meniscus to the underlying bone, thereby maintaining their position within the joint. They are the ligamentous zone, uncalcified fibrocartilage, calcified fibrocartilage, and bone. Coronary ligaments run along the periphery of each meniscus, providing an additional attachment to the tibial plateau [3]. It is here in the periphery of each meniscus that the rest of the innervation is found, with no innervation present in the inner one-third of the tissue. Large nerve fibers run circumferentially along the tissue, while smaller fibers are positioned radially. The outer periphery of each meniscus is also covered by the synovial membrane, which imparts vasculature, and contains a highly fibrous matrix while the inner portion displays characteristics like those of hyaline articular cartilage, devoid of vasculature and innervation.

1.1.2   DEVELOPMENT OF THE KNEE MENISCUS

When first formed in the body, both the medial and lateral menisci are completely vascular. This widespread vascularity diminishes rapidly from gestation to birth and then more gradually to adulthood, when it is estimated that 10-25% of the lateral meniscus and 10-30% of the medial meniscus contains blood vessels. This vascularity is confined to the outer periphery of the meniscus [4].

Understanding meniscus development can greatly enhance tissue engineering efforts by providing a basis for evaluating engineered construct maturation *in vitro* and *in vivo*. A 1983 study on the developing meniscus revealed that from early in gestation to 11 years after birth vasculature as well as cellularity changes dramatically, but contact area between the tissue and bones remains constant [5]. Specifically, at 3.5 months gestation, the meniscus has little extracellular matrix but high cellularity and vascularity. The cells at this stage are similar to each other in that they have a high nucleus to cytoplasmic ratio, but are more compacted on the periphery of the tissue. By around 6 months' gestation, the cruciate ligaments are more defined and the collagen network is fairly well organized circumferentially, including some radial tie fibers. At this stage the meniscus is still completely vascularized. At 7.5 months gestation, the synovial membrane covering the meniscus is around three to four cell layers thick, the collagen organization is more pronounced, and the nuclear to cytoplasmic ratio of the cells has diminished. As the fetus grows larger and approaches 9 months gestation, the meniscus within the joint maintains a relatively constant ratio of contact area with the tibial plateau, correlating with an increase in extracellular matrix around the meniscal cells [5]. Therefore, even before birth, the meniscus undergoes drastic changes in terms of its cellularity, vascularity, and size, which may be useful to mimic in tissue engineered constructs.

After birth, the meniscus continues to grow along with the joint, and the collagen organization changes to accommodate biomechanical loading. At 3 months after birth, vasculature can be identified throughout the meniscus but is more concentrated at the periphery. Vasculature in the inner one-third of the meniscus continues to diminish, and is almost completely gone by 9 months. As the tibia and femur develop, the meniscus continues to increase in collagen organization and size in concert with the increasing femoral and tibial surfaces. From 3 to 11 years, the synovial membrane decreases in thickness to one to two cell layers, and the collagen organization within the meniscus develops to contain not only circumferentially and radially oriented fibers but also vertically oriented fibers. The vasculature during this period continues to recede to the periphery of the tissue (see Figure 1.4) [5].

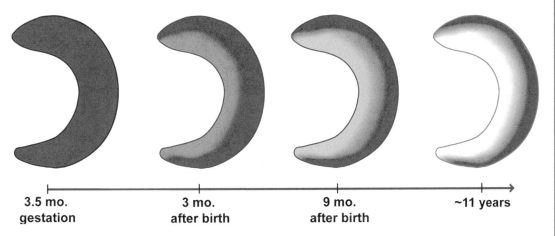

Figure 1.4: Decrease in meniscus vascularity during development. Before birth, the developing meniscus is completely vascularized. After birth, this vascularity begins to recede rapidly toward the outer periphery during the first 9 months, and then more gradually until approximately 11 years. At this stage, the inner one third of the meniscus is completely avascular.

Through adulthood, the meniscus decreases in vascularity and cellularity, eventually becoming avascular in the inner one-third of the tissue. This change in vascularity is directly related to a 20 kDa portion of the C-terminal region of collagen type XVIII called endostatin, which acts as an inhibitor of vascular in-growth through inactivation of vascular endothelial growth factor [6, 7]. Early in development collagen type XVIII is homogeneously distributed throughout the meniscus, but as the aging process continues levels increase in the inner two-thirds of the tissue and decrease in the outer one-third, creating a favorable environment for vascularization in the peripheral meniscus [6]. During early maturation (less than 20 years of age), the rate of proteoglycan production in the meniscus is 1-5 mM sulfate per hour per milligram of DNA and then gradually decreases with aging to around 1/20 of its initial rate [8]. The adult meniscus has a high degree of collagen organization allowing for specialized load transmission from the curved femoral condyles to the flat tibial plateau (see Section 1.3).

As the body moves through adulthood and begins to age, the meniscus undergoes degenerative changes. Collagen concentration increases from birth to 30 years and allows for the creation of a highly organized matrix. It then reaches a plateau from 30 to 80 years of age, and finally begins to decline [9, 10]. Around this time of decline there is an observed increase in ratio of chondroitin-6-

sulfate to chondroitin-4-sulfate, and an increase in keratin sulfate to chondroitin sulfate, characteristics also seen in hyaline cartilage aging [11]. These degenerative changes may increase the risk of the meniscus to become injured (more discussion in Section 2.1).

CONCEPTS

The knee meniscus is comprised of two semi-lunar, wedge-shaped tissues that act to stabilize, absorb shock, bear load, and transfer stresses within the joint. During articulation of the bones, these tissues are held in place by a network of ligaments. Direct ligamentous attachments that integrate the meniscus to the tibial plateau are made by the horns of the meniscus, which contain both cartilaginous and calcified regions. During the initial stages of meniscal development, the tissue is completely vascularized, but with maturation blood vessels are restricted to the outer periphery. Marked differences in cell density and collagen organization can be seen as the tissue matures, as the meniscus contains condensed cells with little extracellular matrix early on, and fewer cells surrounded by abundant organized matrix in adulthood. Meniscal aging is apparent through a decline in collagen and an increase in chondroitin-6-sulfate and keratin sulfate relative abundances.

1.2   BIOCHEMICAL COMPOSITION, STRUCTURE, AND FUNCTION

1.2.1   REGIONAL VARIATION

While the meniscus contains blood vessels and nerves, these are only found peripherally in the tissue, and therefore the meniscus is generally considered in terms of two regions. The vascularized and innervated (red) region is located exclusively in the outer periphery, and the non-vascularized (white) region makes up the inner portion of the tissue (see Figure 1.5) [9, 12, 13]. These two regions are joined together by a transitional region called the red-white region, which exhibits both red and white properties. The capacity for a region to self-repair correlates directly with the amount of vasculature present, giving the red region the highest regenerative potential. The red and white regions also differ greatly in terms of biochemical content, mechanical properties, and cell type.

1.2.2   BIOCHEMICAL CONTENT

Overall, the meniscus is composed of approximately 70% water and 30% organic matter [3]. Of the organic matter, 75% is collagen. Although collagen is present throughout the meniscus, different types are prevalent in different regions. In the red region of the meniscus, collagen type I is the main collagen present while in the white region, both collagens type I and II are in abundance [3, 12]. Other collagens present in the meniscus are types III, IV, V, VI, and XVIII but to a much smaller degree than types I and II [6, 12]. The outer portion of the meniscus is 80% collagen by dry weight and is almost exclusively type I, with less than 1% of other collagen types [14]. In contrast, the inner portion of the meniscus is 70% collagen by dry weight. Of this collagen, 60% is type II and 40% is type I [14]. Therefore, the outer portion of the meniscus is more fibrous and the inner portion of the meniscus, containing collagen type II, has some hyaline cartilage-like properties.

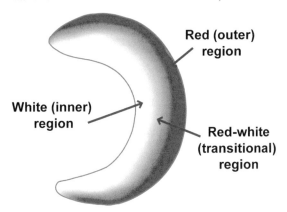

Figure 1.5: **Locations of the three regions of the meniscus.** Vascularity defines regions radially in the meniscus. Closest to the synovial membrane is the red (outer) region, which is highly vascularized. Moving toward the center of the joint space, blood vessels become more sparse in the red-white (transitional) region, and are absent in the white (inner) region.

As the largest fraction of the extracellular matrix, collagen has an important role in the functionality of the meniscus. Being a fibrillar protein, collagen type I is able to confer various types of mechanical integrity based on its structural organization. The alignment of collagen fibers in the meniscus varies from being mostly random within the superficial and lamellar layers, to oriented circumferentially in the deep layer and with radially oriented "tie" fibers present throughout [12]. This alignment allows for the meniscus to withstand hoop stresses generated by normal loading of the tissue.

While the overwhelming majority of fibers in the meniscus are collagens, elastin has also been found in the matrix, though it comprises 1% or less of the dry weight. The presence of even such a small degree of elastin is thought to provide resiliency to the tissue, as it is known for being able to recover its original shape after withstanding large strains. It has been proposed that elastin interacts directly with the collagen network during loading to impart resiliency to the matrix [15].

The remaining 25% of the organic matter in the human meniscus is made up of proteoglycans (∼15%), cellular DNA (∼2%), and adhesion glycoproteins (<1%) [12, 16]. This breakdown can vary regionally within the meniscus. Proteoglycans are molecules consisting of a core protein that is decorated with glycosaminoglycans (GAGs), and are commonly classified based on the GAGs present. Of the GAGs that are found in the meniscus, 40% are chondroitin-6-sulfate, 10-20% are chondroitin-4-sulfate, 20-30% are dermatan sulfate, and 15% are keratin sulfate [16]. GAGs are negatively charged and therefore play a central role in attracting water into the tissue, imparting both hydration and compressive stiffness. Due to the need for compressive integrity, cells from the inner two-thirds of the meniscus produce more proteoglycans than the outer one-third [17, 18]. Biglycan, which is theorized to protect cells during loading, is at its highest concentration in the

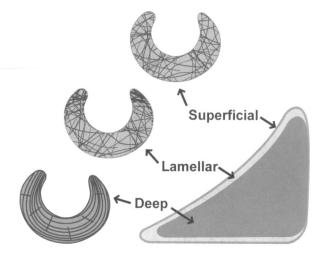

Figure 1.6: **Collagen orientation within the meniscus.** The meniscus displays different collagen organization with depth in the tissue. Here, a vertical cross-section of the meniscus is labeled with the three meniscus zones: superficial, lamellar, and deep. Collagen fibers are randomly oriented in the superficial zone and throughout most of the lamellar zone. Some radial fibers can be detected in the posterior and anterior horns of the lamellar layer. The deep layer is characterized by circumferentially oriented fibers, with some radial fibers dispersed throughout.

inner one-third of the meniscus [17]. In addition, decorin, which helps collagen fibril organization, is found mostly in the outer one-third of the tissue where collagen organization is highest [17]. Due to its wedge shape, the highest compressive loading on the meniscus is borne by the inner portion, while the outer portion experiences a tensile hoop stress (more discussion ensues in Section 1.3). The spatial organization of proteoglycans within the meniscus therefore allows the tissue and the cells within it to withstand compressive loading and to organize collagen fibrils to bear tensile loads.

Adhesion glycoproteins are a specialized class of molecules that aid in binding matrix molecules to one another and to cells. Within the meniscus type VI collagen, fibronectin, and thrombospondin have been identified [9]. All of these molecules contain the Arginine-Glycine-Aspartic acid (RGD) amino acid sequence which aids in cell attachment and which allows for cellular and extracellular matrix connections.

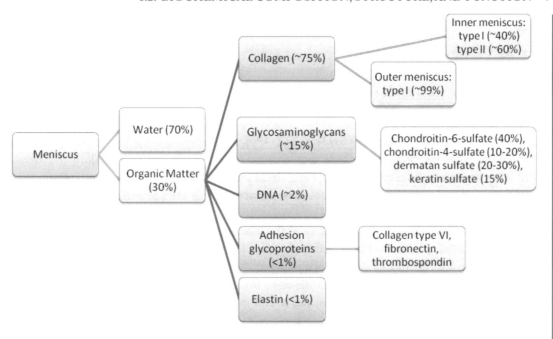

Figure 1.7: **Schematic diagram of meniscus biochemistry.** A hierarchical structure of the overall contents of the meniscus is shown in which water is the largest component. Within the solid fraction, collagens comprise the majority, followed by glycosaminoglycans. DNA, adhesion glycoproteins and elastin form a small fraction of organic matter. Variations in meniscus composition can be observed in different locations within the tissue.

CONCEPTS

Water is the main component of the knee meniscus, comprising 70% of the total wet weight of the tissue. The dry weight contains 75% collagen, mainly types I and II, and 25% proteoglycans, cells, and adhesion glycoproteins. The main GAG present in the meniscus is chondroitin sulfate (50-60%), followed by dermatan sulfate (20-30%), and keratin sulfate (15%). Collagen is preferentially organized in the circumferential direction, with radial fibers dispersed throughout in order to bear tensile loads generated from joint movement. Given this general collagen organization, clear differences can be seen in the biochemical makeup of the outer and inner portions of the meniscus. In the outer portion of the meniscus collagen type I is dominant, but in the inner portion type II collagen is slightly more prevalent than type I. Proteoglycans are also more abundant in the inner portion of the meniscus, allowing for high compressive loads to be borne. Resiliency of the knee meniscus is imparted to the tissue through the slight presence of elastin (<1 % of dry weight). Cells are dispersed throughout the matrix and are anchored to it via adhesion glycoproteins which contain

RGD peptides. Biochemical composition of the knee meniscus varies regionally, and allows for the specialized function of the tissue.

1.3   BIOMECHANICAL PROPERTIES AND EVALUATION TECHNIQUES

1.3.1   GEOMETRICAL CONSIDERATIONS

Because the main functions of the knee meniscus are load transmission and stability, this tissue must withstand many different forces including shear, tension, and compression. The structure and composition of each semicircular meniscus is well-suited to this task, as evidenced by its unique biomechanical properties.

On a macroscopic level, the geometry of the meniscus gives the first indication of its function. The meniscus is both semilunar and wedge-shaped. In terms of its semilunar geometry (as shown in Figure 1.8), the medial and lateral menisci can be measured anteroposteriorly (lengthwise), and mediolaterally (widthwise). Typical dimensions for the medial meniscus are 40.5-45.5 mm in length, and 27 mm in width [19, 20]. For the lateral meniscus, length and width dimensions are typically 32.4-35.7 mm and 26.6-29.3 mm, respectively [19, 20]. The circumferential dimension for the medial meniscus is approximately 90-110 mm, while for the lateral meniscus it is slightly shorter (approximately 80-100 mm) [19].

As the tissue is wedge-shaped, thick on the outer periphery and thin toward the middle of the joint, it is ideally suited to stabilize the femoral head as it articulates with the tibial plateau by increasing congruency between the two surfaces. Additionally, as load is applied from the femur to the tibia, the meniscus draws upon its unique shape to deform radially, thereby bearing some of the load that would otherwise be transmitted to the tibial cartilage. The radial displacement, opposed by posterior and anterior attachments on the tibial plate, results in a hoop stress in the tissue.

1.3.2   NORMAL LOADING CONDITIONS

During normal activities such as walking or ascending stairs, the knee joint experiences loads of 2.7-4.9 times body weight [21]. Overall, it is estimated that the knee meniscus bears anywhere from 45% to 75% of this total joint load, varying with degree of joint flexion, animal model, as well as health of the tissue [22]. As the knee flexes, the contact area between the bones in the joint decreases by 4% for every 30°, accounting for some of the variability in load-bearing capacity of the meniscus [23]. It has been shown that at full extension, the lateral meniscus bears almost all of the load on the lateral side, while the medial meniscus bears about 50% of the medial load [24]. The meniscus not only acts to increase congruence in the joint, it also acts as a spacer creating 1 mm of space between most of the articulating femoral and tibial surfaces and allowing only 10% of these surfaces to contact [24]. Without the meniscus tissue, the support of the femoral condyles is dramatically reduced and the joint force is concentrated on the hyaline cartilages, increasing the stress on the tissue to 2 to 3 times higher than normal [25]. It is therefore evident that the load-bearing capacity of this structure and

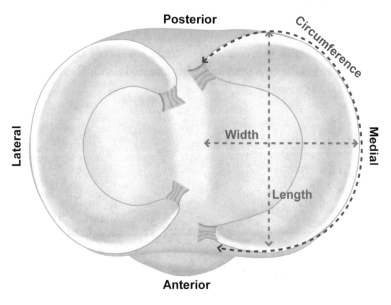

Figure 1.8: **Geometrical measurements of the meniscus.** Typical circumferential measurements for the human meniscus range from 80-100 mm and are shorter for the lateral meniscus than for the medial meniscus. Lengths and widths range from 32.4-35.7 mm and 26.6-29.3 mm, respectively.

its role in protecting the hyaline cartilage surfaces of the femur and tibia act to prevent joint injury. Both geometry and anatomical anchors play an important role in the stabilizing, load-bearing, and protective functions of the meniscus.

1.3.3   SHOCK ABSORPTION
The meniscus also plays a distinct role in absorbing shock within the joint. Studies on the bovine meniscus have shown that this tissue has 1/2 the stiffness and 1/10 the permeability of hyaline cartilage [26, 27]. Additionally, as the collagen fibers within the meniscus have varying diameters, they are suited to absorb a variety of different frequencies [28]. These features make it easier for the meniscus to absorb shock and deform in response to joint movement.

1.3.4   COLLAGEN ORGANIZATION
On a microscopic level an even more refined architecture can be distinguished that allows these specialized functions to be realized. As discussed previously, throughout the developmental process the collagen matrix in the meniscus becomes increasingly organized. This organization varies with depth in the tissue, imparting both tensile stiffness and resistance to splitting [29, 30]. Collagen orientation can be considered in three layers: superficial, lamellar, and deep, which describe the tissue from surface to core. As shown in Figure 1.9, collagen fibers are amorphous in the superior superficial

layer, but are more radially oriented in the inferior superficial layer, closest to the tibial plateau [31]. Amorphous collagen organization persists through the lamellar layer, but is distinguished from the superficial layer in that it contains short, radially oriented fibers only at the posterior and anterior horns [30]. In the deep layer, collagen is predominantly oriented circumferentially, with a few radially oriented fibers [9, 28, 31, 32].

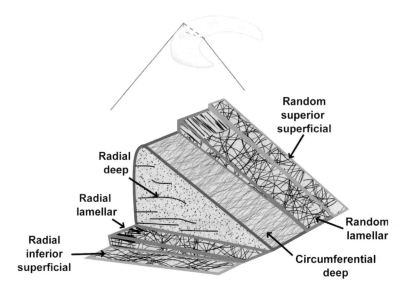

Figure 1.9: Collagen architecture of the knee meniscus. From core to surface, collagen arrangement changes from structured to unstructured. Collagen orientations in the meniscus are of three main types: circumferential, radial, and random. Circumferential fibers are the most abundant in the tissue and are found in the deep zone. Radial fibers are dispersed throughout the deep zone and are present on the periphery and at the horns of the meniscus in the lamellar zone. Despite the presence of radial fibers, random fiber orientation dominates the lamellar zone. In the superficial zone, fiber orientation is typically random in the superior region and more radially oriented in the inferior region.

As described previously, during normal loading conditions the femur presses down on the meniscus, creating radial displacement that is opposed by anterior and posterior anchors. This displacement is translated within the tissue to hoop stresses, radial tension, shear, and compression which are borne by the special organization of collagen fibers and proteoglycans, as shown in Figure 1.10. In the superficial and lamellar layers, amorphous and radial collagen fibers act to resist mediolateral splitting of the meniscus and in the deep zone circumferentially oriented fibers work in tension as a result of hoop stresses [10]. Shear forces generated by the femur deforming the tissue are opposed by matrix molecule interactions, and negatively charged proteoglycans in the meniscus impart compressive integrity by resisting fluid loss [12].

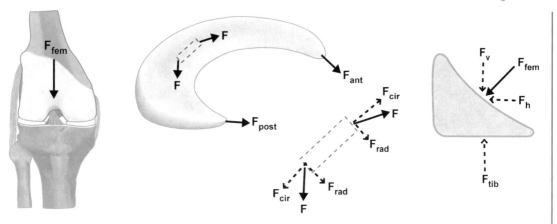

Figure 1.10: **Free body diagram of forces acting on the meniscus during loading.** As the femur presses down on the meniscus during normal loading, the meniscus deforms radially but is anchored by its anterior and posterior horns (F_{ant} and F_{post}). During loading, tensile, compressive, and shear forces are generated. A tensile hoop stress (F_{cir}) results from radial deformation, while vertical (F_v) and horizontal (F_h) forces result from the femur pressing on the curved superior surface of the tissue. A radial reaction force (F_{rad}) balances the femoral horizontal force (F_h).

1.3.5   BIPHASIC BEHAVIOR

Due to its makeup, the meniscus is considered biomechanically as a biphasic tissue. The first phase consists of the porous and permeable collagen and proteoglycan solid matrix, while the second phase is made up of water and salts that are present throughout the matrix [26, 33]. While the solid matrix makes up only about 30% of the total tissue, it is the interplay between the solid and fluid phases that imparts viscoelastic properties to the meniscus (see Figure 1.11) [34]. As a viscoelastic tissue, the mechanical behavior of the meniscus depends on both the magnitude and rate of loading. Frictional drag is produced by fluid being forced from the tissue during load application, producing creep and stress-relaxation responses [26]. When subjected to a constant force or stress applied suddenly as a step, the meniscus displays elastic-like properties immediately after loading. This initial behavior is controlled by the hydrostatic pressure developed in the interstitial fluid portion of the tissue. After this initial phase, still under constant stress, the tissue continues to deform, but at a slower rate as the fluid phase is expelled from the matrix, with the solid matrix resisting more of the load. This deformational behavior under a constant, step load is called the creep response [33].

A similar behavior can be observed when a step strain or displacement is placed on meniscal tissue. Initially, the solid matrix responds elastically by creating a reaction force that is linearly related to the applied displacement. Over time, this reaction force diminishes exponentially as the fluid is expelled from the matrix and the load is shared by both fluid and solid, until eventually only the solid matrix supports the applied load. This behavior is called stress-relaxation [33]. Following

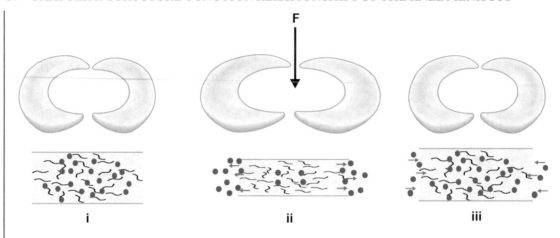

Figure 1.11: Biphasic behavior of meniscal tissue. (i) GAGs (black lines) and water (blue dots) coexist in the matrix, (ii) as the meniscus is loaded, water is forced from the matrix, and (iii) when the load is released, the negatively charged GAGs attract water back into the matrix, rehydrating the tissue.

load removal within the joint, the fluid that was expelled during loading is able to rehydrate the tissue, initiated by the negatively charged proteoglycans in the matrix, resulting in recovery behavior. This fluid flow also functions to transport nutrition throughout the tissue and surrounding hyaline cartilage, remove waste, and provide lubrication [32]. Therefore, the mechanical behavior of the meniscus is not only vital to ensure proper load distribution, but is also instrumental in the overall health and lubrication of the joint.

1.3.6   BIOMECHANICAL EVALUATION

A number of mechanical tests are used to quantify just how mechanically robust this tissue is when subjected to tensile, compressive, and shear loading. Due to the variation in collagen alignment and the asymmetrical shape of the meniscus, a complete picture of the mechanical properties of the meniscus must consider specimens that vary spatially within the tissue and are oriented along and perpendicular to the preferred collagen alignment. Figure 1.12 details the various directions and regions that are important in meniscus characterization. The most common methods used to characterize the mechanical properties of meniscal tissue are tensile and compressive tests. It is important to note that as the availability of human tissue is limited, some mechanical characterization data are only available for other animals such as the cow, pig, or sheep.

1.3.7   TENSION

For tensile testing, tissue can be harvested from the meniscus at a prescribed depth perpendicular to or parallel with the circumference of the meniscus. It is crucial to maintain consistency amongst samples with regard to position and orientation because the meniscus is known to have anisotropic

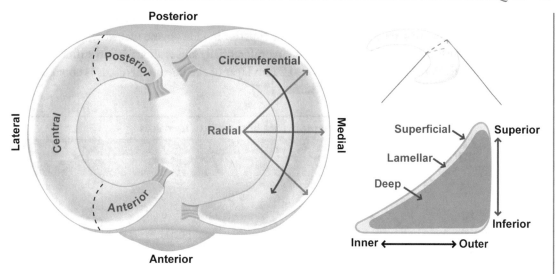

Figure 1.12: **Regions and directions of the meniscus.** Because meniscal properties vary with location in the tissue, defining different regions, depth zones, and directions is useful. The meniscus can be divided into anterior, central, and posterior regions based on location in the joint, and into superficial, lamellar, and deep zones that vary with depth in the tissue. Directions such as inner and outer, as well as inferior and superior describe the tissue from the center of the joint to its periphery, and from the tibial plateau to the femoral condyle, respectively. Also of importance are the circumferential and radial directions, which indicate that the meniscus is semi-circular in nature.

properties [29, 32]. Tensile tests are usually performed using a constant strain rate of 0.005 sec^{-1}, and the samples may or may not be preconditioned [30, 32]. This small strain rate is used to minimize the effects of frictional drag from interstitial fluid flow out of the solid matrix, and to neglect the viscoelastic properties of the collagen and proteoglycan matrix [30].

When considering specimens from the anterior, central, and posterior meniscus it can be seen that the circumferential Young's modulus varies spatially, and that the lateral meniscus has a higher average tensile circumferential modulus than the medial meniscus (see Table 1.1) [32]. When testing the bovine meniscus in the radial direction, its modulus is highest closest to the posterior region of the meniscus, and decreases moving toward the anterior horn [30]. There is evidence that the tie fibers in the posterior region of the bovine meniscus are closely packed and form sheets, which may explain the higher modulus found there [30]. In tension, the properties of the meniscus vary from being isotropic on its surface, to anisotropic in deeper layers due to the variation in collagen fiber alignment. These collagen fibers provide the tissue a robust tensile stiffness of up to 300 MPa [32]. In the superficial layer, the tissue fails at high stresses and low strains with no preferred direction. In the deeper zones, the tensile modulus in the circumferential direction can be 3 to 10 fold higher than in the radial direction, owing to the abundance of circumferential collagen fibers relative to radial fibers [30, 32]. Comparing tensile properties of the superior medial bovine meniscus, the layer 0.6-1 mm from the surface has the highest stiffness circumferentially and is about 4-fold stiffer than the layer 0-0.2 mm from the surface [27]. This 0-0.2 mm layer of the bovine meniscus is about 3-fold stiffer than the layer 1.4-1.8 mm from the surface [27].

Table 1.1: Tensile properties of the native meniscus

Region	Direction	Animal	Stiffness (± SD; MPa)	Reference
Meniscus	Circumferential	Cow		32
Superior			59.8	
Deep			198.4	
Inferior			138	
Lateral	Circumferential	Human		32
Anterior			159.07 ± 47.4	
Central			228.79 ± 51.4	
Posterior			294.14 ± 90.4	
Medial	Circumferential	Human		32
Anterior			159.58 ± 26.2	
Central			93.18 ± 52.4	
Posterior			110.23 ± 40.7	
Medial (from sup. surface)	Circumferential	Cow		27
0–0.2 mm			48.3 ± 29.2	
0.6–1 mm			198.4 ± 87.5	
1.4–1.8 mm			139 ± 79.2	
Meniscus	Radial	Cow		32
Superior			59.8	
Deep			2.8	
Inferior			4.6	
Medial	Radial	Cow		30
Anterior			10-20	
Central			20-40	
Posterior			20-70	
Medial (from sup. surface)	Radial	Cow		27
0–0.2 mm			71.4 ± 41.6	
0.6–1 mm			2.8 ± 1.2	
1.4–1.8 mm			4.6 ± 2.1	

1.3.8   COMPRESSION

Methods for compressive testing of meniscal tissue include confined or unconfined compression and creep indentation [27], [35]–[37]. The creep indentation apparatus is depicted in Figure 1.13.

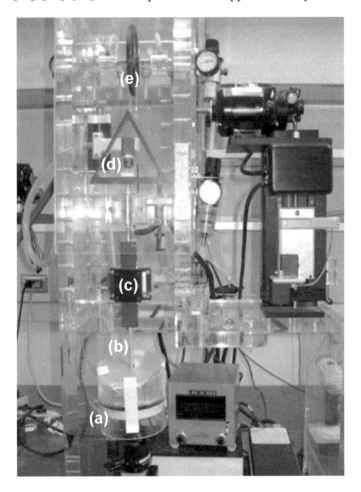

Figure 1.13: **Creep indentation apparatus.** To test a specimen using creep indentation, it is submerged in a buffer solution on the sample mounting stage (a) which is adjusted such that the specimen surface is directly under and perpendicular to the indentation tip (b). An LVDT (c) provides positional feedback to a computer to detect when equilibrium conditions are met after the system receives a step load from the loading stage (d). The entire system is suspended by a pulley (e) which relies on frictionless bearing movement provided by compressed air.

Both compressive testing and creep indentation can yield the aggregate modulus and permeability, but creep indentation can additionally allow for the calculation of the Poisson's ratio, and thus, shear modulus of the tissue. This type of testing has shown that different regions in the meniscus have varying compressive properties, which is a result of their biochemical makeup and organization (see Table 1.2). Using the creep indentation apparatus, it has been shown that the aggregate

Table 1.2: Compressive properties and permeability of the native meniscus using creep indentation or confined compression

Region	Animal	Aggregate modulus ($\pm$ SD; MPa)	Permeability ($\pm$ SD; 10^{-15} m^4 N^{-1} s^{-1})	Reference
Medial superior	Human	Creep indentation		37
Anterior		0.15 $\pm$ 0.03	1.84 $\pm$ 0.64	
Central		0.10 $\pm$ 0.03	1.54 $\pm$ 0.71	
Posterior		0.11 $\pm$ 0.02	2.74 $\pm$ 2.49	
Medial inferior				37
Anterior		0.16 $\pm$ 0.05	1.71 $\pm$ 0.48	
Central		0.11 $\pm$ 0.04	1.54 $\pm$ 0.49	
Posterior		0.09 $\pm$ 0.03	1.32 $\pm$ 0.61	
Medial superior	Cow			37
Anterior		0.21 $\pm$ 0.06	6.22 $\pm$ 2.55	
Central		0.14 $\pm$ 0.05	5.73 $\pm$ 6.19	
Posterior		0.11 $\pm$ 0.04	4.73 $\pm$ 2.56	
Medial inferior				37
Anterior		0.16 $\pm$ 0.06	5.79 $\pm$ 4.31	
Central		0.11 $\pm$ 0.03	5.65 $\pm$ 4.13	
Posterior		0.13 $\pm$ 0.06	5.40 $\pm$ 5.36	
Medial superficial	Cow	Confined compression		27
Anterior		0.39 $\pm$ 0.11	0.76 $\pm$ 0.47	
Central-anterior		0.42 $\pm$ 0.07	0.83 $\pm$ 0.39	
Central-posterior		0.37 $\pm$ 0.08	0.78 $\pm$ 0.38	
Posterior		0.44 $\pm$ 0.11	0.63 $\pm$ 0.47	
Medial deep				27
Anterior		0.49 $\pm$ 0.04	0.91 $\pm$ 0.52	
Central-anterior		0.41 $\pm$ 0.05	0.86 $\pm$ 0.51	
Central-posterior		0.38 $\pm$ 0.09	1.03 $\pm$ 0.58	
Posterior		0.38 $\pm$ 0.04	0.74 $\pm$ 0.14	

modulus of the human meniscus is greatest in the anterior region of the meniscus (around 150 kPa), as compared to the central and posterior regions (around 100 kPa) [37]. Also notable is that the permeability and shear modulus measured within the meniscus are relatively constant amongst all regions [37]. In unconfined compression at 20% strain, the meniscus again displays anisotropic behavior with the highest compressive Young's modulus in the vertical direction being twice as high as in the circumferential and radial directions [36]. This high stiffness in the vertical direction may be attributed to proteoglycans in the matrix of the tissue resisting fluid loss, thereby opposing the vertical force [35, 36]. The compressive integrity of the meniscus allows for axial loading from the femur to be resisted, and because of the geometry of the tissue, some of this vertical loading is translated into circumferential, radial, and shear stresses.

1.3.9   SHEAR

For specifically testing the meniscus in shear, dynamic oscillatory or constant shear strain is applied to the specimen which can measure the dynamic shear modulus as well as the transient shear modulus relaxation function [34]. As mentioned previously, creep indentation can also be used to yield the shear modulus of a meniscus sample, though it is an indirect method of doing so. It has been shown that the dynamic shear modulus of the meniscus is frequency dependent and anisotropic (see Table 1.3). The frequency dependence again points to the viscoelastic nature of the tissue, while the anisotropy of the modulus indicates that collagen organization and interactions between collagen and proteoglycans are central to shear resistance [34]. The normal human meniscus has a shear modulus on the order of 120 kPa at 1.5 Hz and 10% strain [32]. The orientation of collagen fibers within the meniscus has a profound impact on shear modulus at low compressive strains. The shear modulus of a bovine meniscus sample that undergoes a dynamic shear test in the circumferential direction is 20-36% higher than a sample sheared in the radial direction, and depends on the amount of strain applied (up to 10%) [34]. With compressive strains higher than 10%, these differences diminish [34].

CONCEPTS

Because of its geometrical shape and anchoring, the meniscus experiences tension, compression, and shear while bearing load or stabilizing the joint. Collagen and proteoglycans play an important role in imparting robust tensile, compressive, and shear properties to the meniscus and their anisotropic organization is vital to this function. The meniscus is modeled as a biphasic material, displaying viscoelastic behaviors when subjected to step stresses or strains. Biomechanical testing of the tissue reveals that the tissue has a tensile modulus on the order of 100-300 MPa in the circumferential direction, which is 10-fold higher than in the radial direction. It is also well-suited to resist compression axially, given an aggregate modulus of 100-150 kPa. In shear, the tissue exhibits a shear modulus on the order of 120 kPa.

Table 1.3: Shear properties of the native meniscus using creep indentation or dynamic oscillatory strain

Region	Animal	Shear modulus (± SD; MPa)	Reference
Medial superior	Human	Creep indentation	37
Anterior		0.08 ± 0.01	
Central		0.05 ± 0.01	
Posterior		0.05 ± 0.01	
Medial inferior			37
Anterior		0.08 ± 0.02	
Central		0.06 ± 0.02	
Posterior		0.05 ± 0.01	
Medial superior	Cow		37
Anterior		0.11 ± 0.03	
Central		0.08 ± 0.02	
Posterior		0.06 ± 0.02	
Medial inferior			37
Anterior		0.08 ± 0.03	
Central		0.06 ± 0.02	
Posterior		0.07 ± 0.03	
Medial	Cow	Dynamic (10% strain, 10 rad/s)	34
Axial		0.067 ± 0.024	
Circumferential		0.087 ± 0.023	
Radial		0.061 ± 0.028	

1.4   CELL TYPES

1.4.1   CELL CLASSIFICATION

Development of the meniscus begins with the condensation of a vast number of cells that are largely indistinguishable from one another. After the tissue has matured, however, the cells in the different meniscal layers are morphologically distinct. In the superficial layers of the meniscus, cells appear oval and fusiform, similar to fibroblasts [38]–[40]. In the deeper zones, however, cells are found to be more rounded in nature which is more similar to chondrocyte morphology [1], [38]–[40]. These variations have made the classification of these cells difficult. Researchers have used various terms to describe them including fibroblasts, fibrocytes, chondrocytes, fibrochondrocytes, and meniscus cells [1, 38, 39], [41]–[43]. While the meniscus is made up of predominantly collagen type I, it is natural to expect fibroblast-like cells to inhabit it, and indeed the morphology and gene expression patterns of cells in the meniscus exhibit some fibroblastic characteristics; they also exhibit chondrocyte-like characteristics. These cells cannot be strictly classified as chondrocytes either because instead of exclusively producing collagen type II they are known to produce collagen type I as well. The term meniscal fibrochondrocytes has been used to collectively describe these heterogeneous cells, encompassing both their fibroblastic and chondrocytic natures, and studies have been performed to further characterize them [40]. In this book, we use the terms meniscal fibrochondrocytes and meniscus cells interchangeably.

1.4.2   DIVERSITY OF MENISCUS CELLS

In the rabbit meniscus, as many as four distinct cell types have been identified in various locations in the tissue based on morphology and the presence of gap junctions [38]. In the inner one-third of the tissue, cells with a rounded shape closely resembling chondrocytes have been found, while in the outer portion of the meniscus two cell types that display many cell processes are present [38]. Another cell type is found in the superficial zone of the meniscus and has a spindle-shaped morphology [38]. The cell types in the outer portion contain gap junctions, while cells from the inner portion and superficial zone of the meniscus do not contain these processes [38]. In addition to fibrochondrocytes, endothelial cells are present to maintain the microvasculature of the outer meniscus [44]. These cells are distinct from fibrochondrocytes as they are found only in the lumen of meniscal vasculature.

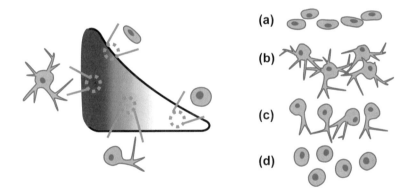

Figure 1.14: **Cell types of the meniscus, distribution, and morphology.** Superficial zone cells are flattened (a), red zone cells display many cell processes (b), red-white zone cells display some cell processes (c), and white zone cells are rounded and chondrocyte-like (d).

1.4.3   CELL SYNTHETIC PROPERTIES

Meniscus cells from all regions work in concert to produce the appropriate proteins needed to maintain healthy tissue. Total collagen synthesis does not vary amongst the regions of the meniscus, but the main types of collagens produced are types I and II. A lower degree of synthetic activity can be detected for collagen types III, IV, V, and VI by these cells [18, 45]. The GAGs produced by meniscal cells are predominantly chondroitin sulfate and, to a lesser degree, keratin sulfate [46].

Comprehensive gene expression profiles for meniscal fibrocartilage as a whole have been compared to hyaline cartilage, identifying common genes and genes specific to each cartilage type [47]. While there are numerous genes that hyaline and meniscal cartilages have in common, there are some genes that are more highly expressed in one cartilage type than the other. When compared to human mesenchymal stem cells (which are considered precursors to cartilage cells), certain genes are more highly expressed in chondrocytes than fibrochondrocytes, and vice versa, as shown in Table 1.4. This type of analysis allows for a better characterization of the tissue and highlights the differences between these two distinct cartilages at a molecular level.

Table 1.4: Genes related to hyaline and meniscal cartilages: Fold-increased expression relative to human mesenchymal stem cells*

Hyaline expression	Meniscus expression	Gene symbol	Gene name
>2	<0.5	IGF2	insulin-like growth factor 2 (somatomedin A)
>2	<0.5	IGL@	immunoglobulin lambda locus
>2	<0.5	RTN4R	reticulon 4 receptor (Nogo receptor)
>2	<0.5	EPHX2	epoxide hydrolase 2, cytoplasmic
>2	<0.5	CREG	cellular repressor of E1A-stimulated genes
>2	<0.5	FLJ13840	Homo sapiens cDNA FLJ13840 fis, clone THYRO1000783
>2	<0.5	BCL7A	B-cell CLL/lymphoma 7A
>2	<0.5	PLA2G2A	phospholipase A2, group IIA (platelets, synovial fluid)
>2	<0.5	CTSC	cathepsin C
>2	<0.5	RBP4	Retinol-binding protein 4, interstitial
>100	~15	COL2A1	collagen, type II, alpha 1
<0.5	>2	HPCAL1	hippocalcin-like 1
<0.5	>2	FLJ20831	hypothetical protein FLJ20831
<0.5	>2	PDLIM1	PDZ and LIM domain 1 (elfin)
<0.5	>2	C1QR	complement component C1q receptor
<0.5	>2	COL1A1	collagen, type I, alpha 1
<0.5	>2	COL1A2	collagen, type I, alpha 2
<0.5	>2	CA12	carbonic anhydrase XII

* From Ochi et al. [47].

1.4.4   REGIONAL VARIATION IN SYNTHETIC PROFILES

Fibrochondrocytes from the inner and outer meniscus have distinct protein synthetic profiles and gene expression patterns, giving rise to the heterogeneous makeup of the meniscus (see Table 1.5). In the inner meniscus, cells display a mostly rounded morphology similar to chondrocytes and stain

Table 1.5: Properties of inner and outer meniscus cells

Property	Associated cell type	Reference
Collagen type II	Inner	39, 45, 50, 53
Aggrecan	Inner	53
NOS2	Inner	53, 54
Collagen type I	Throughout	39, 45, 50, 53
CD34	Outer	51
MMP2	Outer	53
MMP3	Outer	53

positively for α-smooth muscle actin which imparts contractile behavior to the cells [42, 48, 49]. These cells also tend to produce more proteoglycans than the polygonal and fusiform cells of the outer region [18]. Inner region cells can be characterized by higher gene expression and production of collagen type II and aggrecan, as well as negative staining for the cell surface marker CD34, which functions in cell to cell adhesion [45], [50]–[53]. Cells in this region also have high gene expression for nitric oxide synthase (NOS2), which is implemented in nitric oxide production and has been shown to regulate meniscus cell biosynthesis [53, 54].

In contrast, cells of the outer region of the meniscus are characterized by high gene and protein expression of collagen type I, proteases MMP2 and MMP3, and stain positively for the cell surface marker CD34 [38, 45, 51, 53]. The gene profile of these cells is more reminiscent of fibrous tissue due to the high degree of collagen I expression and the expression of proteases which can aid in cellular migration and remodeling of the tissue following injury.

1.4.5   MECHANOSENSITIVITY OF MENISCUS CELLS

Gene expression and protein synthesis of meniscal cells can vary with age and region in the tissue, but are also sensitive to mechanical cues (see Table 1.6). Cells from the inner and outer meniscus are exposed to different cytomechanical environments [55]. Using finite element modeling of the meniscus, it has been predicted that the round, inner meniscus cells experience tensile strains on the order of 7% under normal loading conditions [55]. More elongated, outer meniscus cells, however, are predicted to experience strains ranging from 2–4% [55]. These differences highlight that elongation of a spherical cell will result in a more pronounced shape change than the same deformation of an already elongated cell. When subjected to biaxial strains of 5% *in vitro*, all cells regardless of region increase their total protein synthesis [56]. There is also a marked increase in nitric oxide levels, but this is not accompanied by an upregulation of the NOS2 gene [56]. Therefore, cells from all regions

Table 1.6: Effects of mechanical loading on meniscus cells

Stimulus	Details	Effect	Reference
Normal loading	Finite element model	Strain: inner cell (~7%), outer cell (~2–4%)	55
Biaxial cellular strain, in vitro	Cyclic, 5%, 0.5 Hz, 24 hrs	Increased protein synthesis (larger for outer cells than for inner cells), increased NO levels	56
Static tissue compression	0.1 MPa, 24 hrs	3- to 4-fold decrease in expression of decorin, collagen type I, II; 2- to 3-fold increase in MMP-1 expression	57
Dynamic tissue compression	0.08–0.16 MPa, 0.5 Hz, 24 hrs	2-fold decrease in decorin expression, 4-fold decrease collagen type II expression	57
Dynamic tissue compression	0–0.1 MPa, 0.5 Hz square wave, 24 hrs	Increased NO levels	58
Joint immobilization	*In vivo*	2- to 5-fold decrease in aggrecan expression	59

of the meniscus respond similarly to biaxial strain *in vitro*, but may differ *in vivo* due to cues from various matrix molecules present [56].

Cells from the meniscus change their gene expression profiles in response to different regimens of compressive loading [57]. Under static compressive loading of meniscal tissue, gene expression of decorin and collagen types I and II decrease 3- to 4-fold, while mRNA levels increase 2- to 3-fold for MMP-1 [57]. Under dynamic compressive loading (~1 MPa, 0.5 Hz), decorin expression decreases 2-fold, collagen type II expression decreases 4-fold, and nitric oxide (NO) levels increase [57, 58].

In vivo, joint immobilization at 90° flexion results in a 2- to 5-fold decrease in gene expression for aggrecan, the major proteoglycan of the meniscus, indicating that meniscal cells are dependent upon mechanical cues for normal function [59]. These observations suggest that mechanical stimuli, whether in the form of static or dynamic tensile or compressive stresses, can alter cellular processes to either increase or decrease protein synthesis. Additionally, mechanical cues may be another way for cells to assess the need to create or destroy their surrounding matrix, resulting in macroscopic changes in the tissue.

CONCEPTS

Having both chondrocytic and fibroblastic characteristics, cells of the meniscus have been historically difficult to classify. The term meniscal fibrochondrocytes has been used to highlight the complex nature of the cells. As many as four distinct cell morphologies have been identified in the rabbit meniscus with rounded cells in the inner region, cells with increasing numbers of processes toward the outer region, and spindle-shaped, flattened cells in the superficial region.

Different cell synthetic profiles have been detected based on region within the tissue. Cells of the inner meniscus synthesize both collagen types I and II, while the outer meniscus cells synthesize collagen type I, as well as MMPs 2 and 3 for matrix remodeling [38, 45, 51, 53]. Normal meniscus cell functions are also dependent on mechanical stimulation, as joint immobilization has proved to decrease gene expression for aggrecan 2- to 5-fold [59]. Cells in the meniscus are subject to varying loading conditions, with the outer meniscus cells experiencing tensile strains of 2–4%, and inner meniscus cells experiencing 7%. In response to static compression, meniscus cells decrease gene expression for matrix molecules and increase expression for MMPs, and dynamic compression also decreases collagen and decorin expression but increases nitric oxide levels [57].

CHAPTER 2

Pathophysiology and the Need for Tissue Engineering

2.1 PATHOPHYSIOLOGY AND INJURY

2.1.1 MENISCUS PATHOLOGY

The normal meniscus, as discussed previously, is composed of medial and lateral semilunar, wedge-shaped structures. Various deviations from this normal morphology can occur through abnormal development, disease, degeneration, or traumatic injury. Most commonly, abnormal development results in a discoid meniscus, in which the inner portion of the meniscus extends and the tissue is disc-like in shape. This most often afflicts the lateral meniscus and can be complete, in which the meniscus covers almost the entire articulating surface, or incomplete, covering more surface area than normal (see Figure 2.1). The incidence of discoid abnormality is unclear, but population estimates range from 0.4–5% [60, 61]. Though many cases are thought to be asymptomatic and therefore undiagnosed, some discoid menisci can cause locking of the knee and general knee pain [61].

| Normal | Incomplete discoid | Complete discoid |

Figure 2.1: **Discoid meniscus morphology.** The normal meniscus is semi-lunar in shape, but abnormal development may result in a discoid meniscus. When the inner portion of the tissue covers more area than normal, it is deemed an incomplete discoid meniscus. A complete discoid morphology occurs when the inner portion covers nearly the entire articulating surface.

The meniscus may also be affected by metabolic disease, degeneration, and traumatic injury. Metabolic diseases including calcium pyrophosphate crystal deposition, hemochromatosis, and ochronosis, can cause calcification, gross discoloration, and interference with the overall consistency of the tissue [62, 63]. These disorders heavily compromise the functionality of the meniscus, but cannot be treated locally as they are due to systemic changes in the body. Other types of disorders, such as degeneration and trauma, afflict the meniscus more specifically allowing clinicians to focus treatment on the meniscus itself.

2.1.2   OSTEOARTHRITIS AND MENISCAL DEGENERATION

Little is known about the causes of meniscal degeneration, however with degeneration the meniscus becomes more prone to injury [64, 65]. Osteoarthritis can cause widespread degenerative changes in the meniscus as well as the surrounding hyaline cartilage, and has been implicated in meniscal injury. While in the early 1980's meniscus pathology was found to be weakly correlated with osteoarthritis, researchers have more recently identified meniscal injury in around 75% of patients with symptomatic osteoarthritis [66]–[68].

Osteoarthritis affects the meniscus in multiple ways, resulting in compromised tissue functionality. This disease is known to affect the geometry of the meniscus, causing thickening of the medial posterior and lateral anterior horns, which may in turn affect the biomechanics of the meniscus making it more prone to injury [69]. Osteoarthritic changes in the biochemical makeup of the meniscus may also play a role, as induced osteoarthritis in dogs causes an increase in meniscal water content and changes in GAG content and type over time [70]. This disease may also be associated with calcification of the meniscus, but causality has yet to be confirmed [64]. It has also been shown that severe osteoarthritis causes medial joint space narrowing as the medial meniscus displaces radially, which acts to preserve the tissue at the horns slightly, but overall functionality of the tissue is lost and widespread meniscal degeneration is apparent [71]. Therefore, osteoarthritic degeneration can be an important contributor to meniscal injuries.

2.1.3   TEARS OF THE MENISCUS

Various types of meniscal tears can occur as a result of degeneration and/or trauma. There are four main types of meniscal tears: vertical longitudinal, oblique, radial, and horizontal (see Figure 2.2) [4, 72]. Additionally, there are degenerative (complex) tears which describe an overall fraying of the inner meniscal edge consisting of many different types of tears. The vertical longitudinal tear occurs when the meniscus is split along a circumferential line. These tears can either span the entire thickness of the meniscus vertically (called a bucket-handle tear), or only a portion of it [4, 73]. When a bucket-handle tear occurs, the inner portion of the meniscus is free to intrude into the joint space, causing mechanical opposition to joint movement. The length of vertical longitudinal tears ranges from less than 1 mm to almost the entire circumference of the tissue [4]. Oblique tears are also vertical in nature but extend inward from the inner meniscus in a slanted fashion [4]. These tears are often referred to as parrot beak or flap tears because of their shape. The free end of this type of tear can catch within the joint, inhibiting joint movement. Radial tears are similar to oblique tears but propagate radially, cleaving the circumferential collagen fibers [4]. These tears often exist without any symptoms as their free ends are not as prone to catching within the joint space as other tear geometries. However, radial tears can be especially damaging to the overall function of the tissue if left to propagate. Horizontal tears cut the meniscus into superior and inferior parts. They begin in

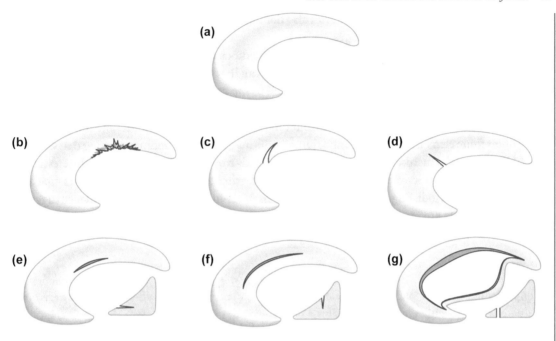

Figure 2.2: **Types of meniscus tears.** The normal meniscus (a) is smooth, wedge-shaped, and semi-circular. Complex (degenerative) tears (b) result in a jagged edge and combine many different types of tears. Oblique tears (c) and radial tears (d) typically propagate from the inner portion of the meniscus to its periphery. Horizontal tears (e) split the tissue into superior and inferior parts and also typically propagate outward. Vertical longitudinal tears (f and g), split the meniscus along the direction of collagen orientation. When a vertical longitudinal tear passes through the tissue's thickness it is called a bucket-handle tear (g).

the inner portion of the meniscus and extend outward, and are often associated with the formation of fluid-filled cysts [4]. These types of tears are thought to be a result of shear forces within the joint and are more common in older patients [4].

2.1.4   EPIDEMIOLOGY OF MENISCUS TEARS

Meniscal tears compromise the overall structural integrity of the joint as well as present symptoms such as locking and catching of the knee, a sensation of giving way, and joint pain [4, 74]. According to one study surveying 1000 patients, meniscus tears occurred more often in the right knee (56.5%) [75]. Of medial meniscus lesions, most (75%) were vertical longitudinal tears and 23% were horizontal tears [75]. In the lateral meniscus the tears are more diverse, with 54% being vertical longitudinal tears and the rest divided amongst oblique and complex pathologies [75]. Overall, meniscal tears affect men more often than women, with 70–80% of meniscus tears occurring in men [4]. Afflicted men are most often 21–30 years of age, whereas this pathology affects women most often between the ages of 11 and 20 [4]. Traumatic injuries also dominate in younger patients, while older patients are more prone to degenerative changes.

Meniscal health is heavily reliant on the ligamentous attachments of the knee. Joint laxity (instability of the joint) as a result of a ruptured ACL can have a profound effect on the meniscus as it has been estimated that the ACL contributes 85% to the restraint of anterior displacement of the femur [76, 77]. Clinically, meniscal injuries are common in patients with torn ACLs, highlighting the co-dependence of the meniscus with surrounding ligaments for normal joint function [4, 78, 79]. Non-linear finite element modeling of knee joints confirms this clinical finding, showing that without the ACL, the medial meniscus is subjected to higher loads from 0° to 30° flexion [80]. Interestingly, though the biomechanics of the knee are altered by tearing an ACL, the types of tears that the meniscus endure are indistinguishable from those of an ACL-intact knee [81]. This evidence suggests that meniscus tears are more frequent when knee stability is compromised, and that meniscal tears follow certain patterns regardless of ligament health.

CONCEPTS

Meniscal abnormalities such as the discoid meniscus may be benign or symptomatic, but are rare, estimated to afflict only as much as 5% of the population. Systemic diseases can change the pigment and consistency of knee meniscus tissue, but cannot be treated locally. Osteoarthritis is an important contributor to the overall health of the meniscus and has been implicated as a cause of meniscal tears and degeneration. The most common injury specific to the knee meniscus is a meniscus tear. Tears can be classified into four different categories based on their geometry: vertical longitudinal, oblique, radial, and horizontal. The most common in both the medial and lateral menisci is the vertical longitudinal tear. Overall, men are affected by meniscus tears 70–80% more often than women, and are usually in their 20's when this occurs [4]. Meniscus tears can come about through degenerative changes or trauma, degeneration being a catalyst in older ages. ACL tears significantly compromise knee stability and increase the likelihood of a meniscal tear.

2.2   THE MENISCUS HEALING PROBLEM

2.2.1   INTRODUCTION

As the meniscus was originally thought to be a vestigial tissue and because surgery on it was difficult, early treatments (prior to the mid-1960s) for meniscal damage were limited to total removal of the tissue, called meniscectomy [82, 83]. As early as 1948 it was shown that meniscectomy causes joint space narrowing, and many studies have subsequently shown that degenerative changes also occur following this procedure [84]–[87]. After more became known about the importance of the meniscus in load distribution and stability within the knee joint, treatments shifted to partial meniscectomy, surgical repair, or transplantation, which are still used today [3, 82, 88, 89].

2.2.2   HEALING IN THE MENISCUS

Given that the meniscus is not a homogeneous tissue, it is particularly difficult for it to self-repair. While lesions or tears that occur in the outer periphery of the tissue can regenerate due to the high degree of vasculature there, damage to the inner non-vascularized portion of the tissue is unable to heal on its own [3, 4, 12, 88], [90]–[92]. Following injury in the vascular portion of the meniscus, the defect site is filled with a fibrin clot which uses proinflammatory factors to recruit blood vessels from the surrounding areas [92]. After this initial response and depending on proximity to abundant blood vessels, fibrous scar tissue can take as little as 10 weeks to form [92, 93]. After a few months, the scar tissue will then mature into tissue with inferior mechanical properties to the native meniscus [92]. This timeline is extended with distance from the peripheral blood supply, and does not occur for injuries in the inner meniscus. For inner meniscus injuries, some reorganization of the matrix may occur due to the changed mechanical environment, but a healing response is absent [94, 95]. Some research has focused on creating vascular access channels from the outer to the inner meniscus to allow healing factors from the blood to reach the damaged white zone, which has helped heal longitudinal tears in the avascular region of dogs and goats, and has reduced symptoms in patients [96]–[99]. As a result, proximity to blood vessels is the best predictor of a meniscal healing response, but the type of healing that takes place does not restore tissue functionality.

2.2.3   CHARACTERISTICS OF REPAIR TISSUE

Though the outer portion of the meniscus may heal to some degree, the new tissue is quite different from native tissue. Repair tissue in the outer portion of the meniscus is distinct from normal meniscal tissue in that it may contain calcified regions, cysts, unattached collagen fragments, and pools of proteoglycans [28]. This is in stark contrast to normal tissue in which the collagen matrix is highly aligned with proteoglycans throughout and no calcification or void spaces. In torn menisci it is twice as common for the tissue to become calcified over time, and this is often found in conjunction with osteoarthritis [100]. Apoptosis is also increased 2-fold in tissue having undergone traumatic injury or degeneration as compared to normal tissue [101].

Functionally, meniscal repair tissue is weaker than normal tissue. Meniscal repair tissue in rabbits has been measured to require around 75% less energy (0.8–0.9 mJ) to fail than normal tissue at 12 weeks post-injury [102]. The strength of this tissue increases only marginally with the use of sutures or fibrin glue to hold the torn edges together, and does not reach normal values [102]. Therefore, even in the region of the meniscus that undergoes some repair, the tissue is either lacking in quantity or insufficient in strength and there is an inhibitory environment for new tissue formation. This evidence points to the need for tissue engineering to develop technologies to heal meniscal injuries and prevent calcification or apoptosis from taking place.

CONCEPTS

Meniscus tissue was originally thought to be vestigial, but upon further investigation it has proved to be an important part of normal knee function and maintenance of healthy tissue. Since this realization, preservation of meniscal tissue is a main priority for reparative therapies. The meniscus is able to heal on its own to a certain degree, but this only occurs when the defect site is in the outer periphery of the meniscus, near vasculature. Even after self-repair has taken place, the scar tissue formed requires 75% less energy to fail than native tissue and may contain calcification, pockets of proteoglycans, and unorganized collagen fibrils. This type of matrix therefore compromises the mechanical integrity of the meniscus as a whole, highlighting the need for tissue engineering to provide an alternative viable solution.

2.3   TISSUE ENGINEERING AND HISTORICAL PERSPECTIVES

2.3.1   DEFINITION OF TISSUE ENGINEERING

Tissue engineering strives to recreate the complex tissues of the body by harnessing biological processes. Using the classic tissue engineering paradigm, cells are seeded onto a scaffold and are subjected to biochemical and/or mechanical stimuli to create a tissue engineered construct ready for implantation. Recently, a new paradigm has emerged which is based on the ability of cells to self-assemble without the use of a scaffold, showing promising results in the field of cartilage regeneration [103, 104]. Tissue from either paradigm may be used to repair damaged areas in the body permanently, thereby eradicating the need for synthetic implantable prostheses that have a limited lifetime of service. This field is especially relevant to the meniscus, which has a limited reparative potential, and its complex biomechanical properties render synthetic replacement materials insufficient.

2.3.2   HISTORICAL PERSPECTIVES

The first use of the term "tissue engineering" was by Y. C. Fung in 1985 in a proposal to the National Science Foundation for the creation of a tissue engineering facility [105]. He presented the idea of tissue engineering as a field to bridge the gap between biology, which studies the single cell, and medicine, which is primarily concerned with the functioning of entire organs [105]. Though his proposal was not accepted, the concept of tissue engineering surfaced again at many different symposia and meetings throughout the late 1980s, prompting much discussion and debate over the exact definition of the term [105]. Beginning in the early 1990s, a steady increase in journal articles can be found containing the term "tissue engineering" indicating its acceptance and popularity throughout the scientific community [105]. Also during this time many different centers for tissue engineering around the world were established, which continue to conduct advanced research in the field [106]. Tissue engineering, though a relatively new concept, has quickly blossomed into a field

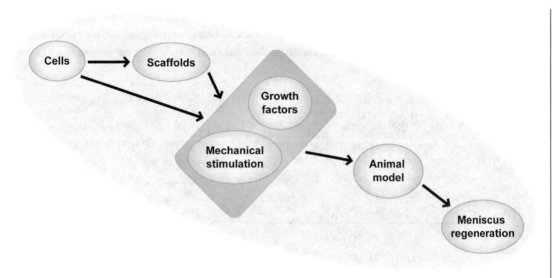

Figure 2.3: **Tissue engineering paradigm.** In the classical tissue engineering paradigm, cells combine with a scaffold and are subjected to a regimen of growth factors and/or mechanical stimulation. The resulting construct is then implanted into an animal model with the expectation that meniscus regeneration will ensue. A scaffoldless paradigm has recently emerged in which cells are seeded at a high density and are conditioned biochemically and/or biomechanically.

with wide application and promise to fulfill Y. C. Fung's vision of bridging the gap between science and clinical implementation.

2.3.3   FUNCTIONAL TISSUE ENGINEERING AND THE MENISCUS

As the field has progressed, so have the standards for success. Functional tissue engineering refers to the emerging effort to not only recreate the biochemical traits of a tissue, but also match native mechanical properties of load bearing tissues [107]–[111]. To do this, researchers rely heavily on tissue characterization research, and use this information to hone in on the most important biochemical and mechanical properties of the tissues they are working to reproduce. For biomechanically active tissues, such as the knee meniscus, it is vital that these characteristics be recapitulated in the engineered version in order that it perform as native tissue and thus be a useful replacement in the body.

Specifically of importance are the tensile and compressive properties of the knee meniscus as well as its geometry, biochemical content, and matrix molecule organization. Attempts at functional tissue engineering of the knee meniscus are ongoing, but key progress has been made in producing similar matrix proteins and geometry to the native tissue [103, 112, 113]. Unfortunately, many tissue engineering studies do not evaluate the mechanical properties of the constructs produced, making functional assessment and comparison to native tissue more difficult. Ideally all functional aspects of a tissue engineered construct should be tested, in order to better understand the potential of each engineering method. A more detailed discussion of the studies in meniscal tissue engineering ensues in Chapter 3.

CONCEPTS

Tissue engineering is a multidisciplinary field which aims to recreate biological tissues using a combination of cells, scaffolds, and biochemical and mechanical stimuli. These tissues may then be used to replace or restore function to missing or damaged elements in the body. A relatively new field, tissue engineering efforts increased in the 1990s with the establishment of research centers around the world. Functional tissue engineering is especially important for the meniscus as it emphasizes mimicking not only tissue biochemistry, but also mechanical and geometric properties. Functional assessment of tissue engineered meniscal constructs is the next step in developing useful meniscal replacement tissue.

CHAPTER 3

Tissue Engineering of the Knee Meniscus

3.1 BIOREACTORS

3.1.1 INTRODUCTION

The meniscus is a mechanically-sensitive tissue that can respond favorably or unfavorably to biomechanical stimuli [114]. It is known that during development, the meniscus experiences a myriad of mechanical stresses that are important for normal maturation. Studies have shown that immobilizing chick embryos results in the complete absence of the meniscus and fusion of the cartilaginous parts in the joint [115]. Additionally, the mature meniscus is also heavily reliant on mechanical stimulation to maintain its health. Joint immobilization of rabbit knees for 8 weeks results in degeneration of the deep zone of the meniscus and reduced tissue permeability [116, 117]. Given the importance of mechanical cues for normal meniscal development and function, it is natural to expect that mechanical stimulation of tissue engineered meniscal replacements would be beneficial. To impart mechanical forces to tissue engineered constructs many different bioreactors have been developed which can apply compression, shear, hydrostatic pressure, vibration, or combinations of these forces in a controlled environment. As the exact type and pattern of optimal mechanical stimulation of the meniscus is not known, various stimulation regimens should be investigated.

There are only a few bioreactors that are specifically designed for stimulation of meniscal tissue, and they comprise a subset of bioreactors designed for stimulation of cartilage constructs. As cartilages in the knee experience similar types of forces, bioreactors generally designed for cartilage stimulation can also be considered applicable to engineering the meniscus. Cartilage bioreactors fall into five different categories: direct compression, hydrostatic pressure, high shear systems, low shear systems, and ultrasound. Other bioreactors use combinations of these in an effort to achieve more tailored results.

3.1.2 DIRECT COMPRESSION

Direct compression bioreactors, sometimes called cyclic strain bioreactors, impart a compressive load to cartilage constructs to simulate normal loading patterns (see Figure 3.1). Typically, strains of 1–10% have been used with frequencies of 0.1–1 Hz. Varying degrees of success have been achieved using this type of stimulation including increases in hydroxyproline content, indicating collagen increases, and [^{35}S] sulfate incorporation, indicating an increase in glycosaminoglycans following the loading regimen (see Table 3.1) [118]–[122]. This type of bioreactor has been used to stimulate meniscal explants, showing an increase in aggrecan gene expression by 108% with dynamic stimulation (2% oscillatory strain, 1 Hz) [123]. Direct compression bioreactors may also be used to

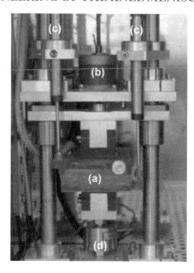

Figure 3.1: **Direct compression bioreactor.** The main component of a direct compression bioreactor is the loading chamber (a), inside of which constructs are placed for stimulation. A stepper motor (b) induces axial deformation to constructs in the load chamber and receives spatial feedback from the LVDTs (c) and force feedback from the load cell (d).

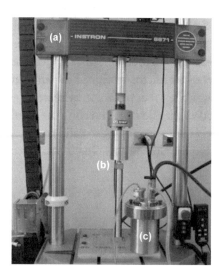

Figure 3.2: **Hydrostatic pressure bioreactor.** An Instron machine (a) drives a hydraulic piston (b) which pressurizes a sealed chamber (c). For stimulation, constructs are sealed in media-filled bags and placed in the water-filled chamber. The chamber is submerged in a water bath and pressurized via the hydraulic piston.

Table 3.1: Summary of direct compression bioreactors

Stimulation	Specimen	Effect	Ref.
3% cyclic strain, 0.1–1 Hz, 6 wks	20×10^6 bovine chondrocytes/ml agarose gel	15–25% increase $[^{35}S]$-sulfate incorporation, 10–35% increase $[^3H]$-proline incorporation	118
10% cyclic strain, 1 Hz, 5 days/wk, 3 hrs/day, 4 wks	60×10^6 bovine chondrocytes/ml agarose gel	1.74% GAG/wet weight 2.64% collagen/wet weight	119
10% cyclic strain, 1 Hz, 5 days/wk, 3 hrs/day, 4 wks	20×10^6 bovine chondrocytes/ml agarose gel	Aggregate modulus: 53 kPa, Young's modulus: 66 kPa, 1.43% GAG/wet weight 1.23% collagen/wet weight	120
10% cyclic strain, 1 Hz, 5 days/wk, 3 hrs/day, 4 wks	60×10^6 bovine chondrocytes/ml agarose gel	No increase in mechanical properties or GAG/wet weight (~1.6%); collagen/wet weight significantly higher (1.86%)	120
1–5% cyclic strain, 1 Hz	Bovine hyaline explant	20–40% increase in $[^{35}S]$-sulfate and $[^3H]$-proline incorporation	121
2% cyclic strain, 1 Hz, 4 hours	Bovine meniscus expants (medial and lateral)	108% increase aggrecan gene expression, no effect on collagen type II expression	123

simulate loading patterns of the meniscus following partial meniscectomy, showing that removing part of the meniscus results in loading patterns that detrimentally affect proteoglycan retention in the tissue [124].

3.1.3 HYDROSTATIC PRESSURE

Static and dynamic hydrostatic pressure bioreactors have also been used to simulate the pressurization of the interstitial fluid or the joint capsule (see Table 3.2). Using this type of loading, constructs or explants are placed in a chamber which is pressurized via a piston to physiologic levels (see Fig. 3.2). Magnitudes of 0.1–15 MPa either applied statically or dynamically at 0.01–1 Hz have been used to stimulate cartilage constructs and have been shown to be effective at increasing mRNA levels responsible for collagen types I and II and aggrecan, and also to reduce tissue breakdown of meniscal explants [117], [125]–[132]. Hydrostatic pressure can be applied to constructs intermittently, in which a period of static culture is interrupted with hydrostatic pressure stimulation, or continuously,

Table 3.2: Summary of hydrostatic pressure bioreactors			
Stimulation	**Specimen**	**Effect**	**Ref.**
10 MPa, 1 Hz, 4 hrs/day, 4 days	10^5 cow wrist cells/cm^2 monolayer	20-fold increase aggrecan mRNA, 9-fold increase collagen type II mRNA	125
1 MPa, 0.5 Hz, 1/14 min (on/off), 4 hrs	Rabbit meniscus explant	Prevents up-regulation of catabolic agents (MMPs, TIMPs) that occurs in un-stimulated controls	117
6.87 MPa, 5/15 s (on/off), for 20 min every 4 hrs, 5 wks	5–7.5x10^6 juvenile horse chondrocytes/cm^3 PGA scaffold	20-fold increase in GAG, 2.4-fold increase in collagen	126
3.45 MPa, 5/15 s (on/off), for 20 min every 4 hrs, 5 wks	2x10^6 juvenile horse chondrocytes/cm^3 PGA scaffold	2-fold increase GAG concentration, no increase in collagen	127
10 MPa, static, 1 hr/day, days 10–14 in culture, 4 wks	5.5x10^6 cow chondrocytes, scaffold-less construct	Aggregate modulus 273 kPa, Young's modulus 1.6 MPa, 6.1% GAG/wet weight, wline 10.6% collagen/wet weight	128
10 MPa, static, 1 hr/day, 1/3 days (on/off), 4 wks	1x10^6 rabbit meniscus cells/ml PLLA scaffold	4-fold increase in collagen production, 3-fold increase in GAG production	130
10 MPa, 1 Hz, 4 hrs/day, 5 days/wk, 8 wks	5.5x10^6 cow chondrocytes, scaffold-less construct	Prevents decrease in GAG content observed in static culture	131
10 MPa, static, 4 hrs/day, 2/1 days (on/off), 1 wk	10^5 temporomandibular joint disc (cow) cells/ml, PGA scaffold	Increased collagen type I gene expression and collagen production	132

in which constructs are cultured within a pressurized chamber. Continuous culture provides the added advantage of minimal construct manipulation, minimizing user variables.

3.1.4   SHEAR

Bioreactors that impart shear forces to constructs can be considered in two classes: high shear systems and low shear systems (see Table 3.3). Both use fluid flow to impart mechanical forces. Two types

Table 3.3: Summary of shear bioreactors

Stimulation	Specimen	Effect	Ref.
Spinner flask, 8 wks	25×10^6 bovine chondrocytes /cm^3 PGA scaffold	60% increase in GAG, 125% increase in collagen, fibrous capsule formation	138
Flow perfusion	12×10^6 ovine chondrocytes/cm^3 PET scaffold	Outer layer of fibroblast-like cells, inner core of meniscus-like cells	135
Parallel plate, 1 dyne/cm^2, 3 days	1.7×10^6 bovine chondrocytes/cm^2, scaffold-less	113% increase in total collagen, 230% increase in collagen type II, Young's modulus 2.28 MPa, ultimate strength 0.81 MPa	139
Rotating wall, 6 wks	127×10^6 bovine chondrocytes/cmp^3, PGA scaffold	GAG levels 68% of native, collagen levels 33% of native, collagen type II crosslinked	140
Rotating wall, 7 wks	5×10^7 rabbit fibrochondrocytes/cm^3, PGA scaffold or agarose	No significant effects over static control	141
Wavy-wall, 4 wks	5×10^6 bovine chondrocytes PGA scaffold	30% increase in cell number and 60% increase in ECM deposition over controls, fibrous outer capsule with type I collagen, inner core with type II collagen and GAG	142

of high shear systems are currently in use: flow perfusion chambers and spinner flasks. In flow perfusion chambers medium flows through the construct, creating shear forces as the fluid makes its way through the pores of the construct [133]–[136]. Another type of high shear environment can be achieved using a spinner flask in which medium is stirred around the constructs, creating high shear as it flows around the boundary of the construct. The main difference between these two approaches is that the perfusion chamber creates shear within the construct and the spinner flask creates shear at the construct's surface. Flow in the perfusion chamber imparts higher shear at the first contact with the construct, and lower shear as the medium exits the construct, which can cause differences in cellular response based on the construct thickness [137]. Spinner flasks, however, only impart shear on the outside of the construct which affects cells at the surface differently than in the center. Despite their differences, both of these bioreactors have been shown to increase collagen and GAG production, while the spinner flask has been especially effective at creating bi-zonal tissues with an outer fibrous capsule [138]. For meniscal tissue engineering, high shear systems such as spinner flasks may be beneficial to create the fibrocartilaginous outer portion while maintaining a more hyaline-like core.

Low shear systems such as parallel plate flow bioreactors and rotating wall bioreactors also impart fluid forces to the constructs, but the forces felt are much lower than those in the high shear systems. A laminar flow bioreactor has been shown to increase collagen type II production and increase tensile properties of cultured chondrocytes, indicating that low shear can be a beneficial way to stimulate cartilage formation *in vitro* [139]. Much work has been done using rotating wall bioreactors in which small amounts of fluid shear are imparted to constructs as they are cultured in a medium-filled chamber that slowly rotates. This environment simulates microgravity by maintaining constructs in a mostly free-floating state. The mechanical environment has shown mixed results in tissue engineering cartilage, in some cases increasing matrix production and in other cases not appreciably improving construct properties [140, 141, 143]. An intermediate form of the rotating wall bioreactor is the wavy-walled bioreactor in which constructs are subjected to a more complicated set of fluid stresses due to the irregular shape of the bioreactor walls [142, 144, 145]. In this type of bioreactor, ECM synthesis is stimulated to varying degrees based on the position of the construct in the chamber [146].

3.1.5   ULTRASOUND

Another type of stimulation that may be used for tissue engineered cartilage is low-intensity ultrasound. The impetus for using ultrasound stimulation came from research done in the late 1990s which showed that ultrasound increased the gene expression for aggrecan in cultured chondrocytes [147]–[149]. This type of stimulation has shown a similar, but less persistent enhancement of ECM formation when compared to the rotating wall bioreactor [150]. Pulsed ultrasound stimulation of cartilage constructs at 1 MHz and 67 mW/cm^2 for 10 minutes per day was shown to be beneficial for cell proliferation and matrix production in a cell-seeded construct for up to 4 weeks, beyond which decay of the construct was observed [150]. The effect of low-intensity ultrasound also has a minimal effect on cartilage construct maturation [151]. Ultrasound therefore may be effective during the cell culture stages of tissue engineering, but has not shown great promise for stimulating cartilage construct maturation.

Table 3.4: Summary of ultrasound bioreactors

Stimulation	Specimen	Effect	Ref.
1 MHz pulsed, 50 or 120 mW/cm^2, 10 min/day, 3 or 5 days	Rat chondrocytes, monolayer	Increase in aggrecan gene expression and [^{3}5S]-sulfate incorporation	148
1 MHz pulsed, 67 mW/cm^2, 10 min/day, 6 days	4x10^6 human chondrocytes/cm^3 scaffold	Increased cell proliferation, beneficial effects last around 28 days (shorter than rotating wall bioreactor)	150
1.5 MHz, 30 mW/cm^2, 20 min/day, 7 days/wk, 12 wks	30x10^6 bovine chondrocytes/cm^2, PLGA scaffold, implanted into nude mice	No effect on accelerating chondrogenesis or maturation of engineered constructs *in vivo*	151

Table 3.5: Summary of combination bioreactors

Stimulation	Specimen	Effect	Ref.
Compression and shear, 0.5 N	Bovine chondrocytes, scaffoldless, immature cartilage-like construct	260% increase in aggrecan gene expression, 310% increase in collagen gene expression at 2 hrs, return to normal after 4 hrs; 4 days increases GAG content	152
Oscillating pin (5% strain, 0.1 Hz), and ball(0.6 Hz, ±60°), 1 hr/day, 3 days	Bovine nasal cartilage explant	Slight increase in COMP gene expression	153
Perfusion and compression, static (10% strain), or dynamic (5% strain, 0.3 Hz), 37 days	Bovine chondrocytes seeded in PGA scaffold	Flat top and bottom, increased GAG retention in construct	136

3.1.6   COMBINATIONS

Limited studies have been performed on combinations of mechanical stimulation. Direct compression has been combined with shear in two different bioreactor systems, emulating physiologic loading patterns and increasing cartilage matrix production [152, 153]. Both bioreactors use a round piece that rotates and imparts a shear force while pressing down on the construct to impart compression. Using an oscillating pin and ball mechanism simulates physiologic compression and shear within a joint and has been shown to increase cartilage oligomeric protein synthesis, which helps to organize cartilage matrix [153]. Another type of combination bioreactor uses fluid shear with cyclic compression and shows higher GAG retention with varying matrix compositions on the surface of the construct versus the core [136]. Combinations of mechanical stimulation can therefore be beneficial to enhancing construct properties beyond that which can be reached using one type of stimulation.

3.1.7   APPLICATION TO MENISCUS ENGINEERING

Based on the research that has been done using these various systems to tissue engineer hyaline cartilage, some projections for their potential use in meniscal tissue engineering can be made. High shear systems may be especially useful for meniscal tissue engineering as they can create bi-zonal tissue, an important attribute of the meniscus. Hydrostatic pressure bioreactors have achieved marked success in increasing mechanical properties of tissue engineered cartilage constructs, a central concern for engineering biomechanically important tissues like the meniscus. Direct compression may be beneficial for meniscal constructs during the later stages of maturation, when mechanical stimulation is known to be important for normal tissue function. Low shear systems could provide stimulation for overall ECM production, but due to the mixed review of this stimulation in articular cartilage engineering it is unclear how beneficial it might be. Ultrasound stimulation is fairly new to cartilage tissue engineering and may be beneficial for stimulating GAG synthesis of cultured cells, but has not shown beneficial effects on construct maturation. Rather than choosing one type of bioreactor, it is likely that some combination of these stimuli will provide the best environment for meniscus tissue regeneration, given the vast number of events that occur during meniscal development.

CONCEPTS

Tissue engineering the knee meniscus has garnered much attention in the scientific community due to the prevalence of meniscus degeneration and defects and the tissue's limited ability to self-repair. As the meniscus is dependent on mechanical stimulation for normal function, bioreactors that simulate physiologic stresses have been developed and may provide the appropriate conditions for meniscal regeneration *in vitro*. Though there are few bioreactors specifically designed for meniscus tissue engineering, a basic understanding of cartilage bioreactors as a whole provides the necessary background for the progression of this field. Five different categories of cartilage bioreactors exist: direct compression, hydrostatic pressure, high shear systems, low shear systems, and ultrasound. Amongst these, direct compression, hydrostatic pressure, and high shear systems seem the most promising for developing tissue mechanically and biochemically similar to the native meniscus.

Combination bioreactors incorporate two or more types of these stimuli, and could allow for more directed formation of the meniscus.

3.2   *IN VITRO* TISSUE ENGINEERING

3.2.1   INTRODUCTION

According to the classic tissue engineering paradigm, the basis of any meniscus tissue engineering attempt are cells and scaffolds. Currently in the meniscus tissue engineering field, the focus has been mostly on characterization of meniscal fibrochondrocytes and their sensitivity to various biochemical stimuli. This information has been used to create various methods for tissue engineering the meniscus that maximize matrix synthesis and increase the biochemical and mechanical relevance of engineered constructs.

3.2.2   CELL SOURCE

Success of tissue engineering strategies relies heavily on the type of cells used and their potential to create enough of the appropriate extracellular matrix molecules. Fibrochondrocytes are well characterized and have been used in many meniscus engineering attempts [18, 40, 42, 43, 46, 154, 155]. While the presence of these cells is optimal for understanding the culture environments and stimuli appropriate for engineering a meniscus-like tissue, they are relatively scarce in the body and would likely be an impractical source for large-scale engineering attempts. This has prompted research into other sources of cells that may present an abundant, autologous or allogenic source.

Other cell types that have been used in tissue engineering the meniscus include chondrocytes, mesenchymal stem cells, and embryonic stem cells. Chondrocytes are nicely suited to meniscal engineering as they can be induced to produce collagen type I through passaging. Their synthetic capacity for collagen and GAGs in tissue engineering is also higher than fibrochondrocytes [113]. Chondrocytes are, however, also relatively scarce in the body and therefore would prove difficult to procure for tissue engineering. Mesenchymal and embryonic stem cells, on the other hand, present a potentially limitless cell source. The difficulty in using these cells lies in understanding the necessary stimuli for directing their differentiation to a fibrochondrocytic lineage. Encouraging results have been obtained using various growth factors including IGF-I, TGF-β1, TGF-β3, BMP2, and BMP6, for directing MSCs from the bone to produce collagen types I and II [156]–[158]. Human embryonic stem cells have also shown promise toward producing fibrocartilaginous tissue when conditioned with transforming growth factors, bone morphogenetic proteins, and when co-cultured with native fibrochondrocytes [159]. These embryonic stem cells, when differentiated and formed into scaffoldless constructs, can produce matrix molecules important for meniscus tissue including collagen types I and II, as well as GAGs [113, 160]. More work needs to be done to solidify the differentiation regimen for both of these cell types, and also to identify MSCs that are sensitive to fibrochondrocytic differentiation.

There are also some cell types that are promising for use in this field, but have yet to be fully investigated for meniscal tissue engineering. These include auricular chondrocytes, dermal fibroblasts, adipocytes, and synovial tissue cells. All of these cell types are advantageous for tissue engineering in that they are easily procured, and have minimal donor site morbidity. Auricular chondrocytes, from the elastic cartilages of the nose and ear, have high proliferative and synthetic capacities and can produce a matrix of collagens I and II that is similar in makeup to meniscal tissue [161]. Recently a subpopulation of dermal fibroblasts has been identified that has the capacity to be chondroinduced through various means including gene transfection and culture on or with various proteins [162]–[164]. This conditioning produces cells that synthesize collagens I, II, and GAGs, which are all important components of the normal meniscus. Like dermal fibroblasts, a subset of adipocytes has also been identified as a cell source for cartilage tissue engineering. Using culture medium containing BMP6 has proved effective for differentiating these cells towards a chondrocytic pathway [165]–[167]. Synovial tissue cells may also prove beneficial for meniscal engineering as they have shown chondrogenic capacity, and an ability to vary production of collagen types I and II in response to the culture environment [168]. All four of these cell sources may provide autologous cell sources for meniscal tissue engineering but more research must be completed to illuminate their potential.

3.2.3   GROWTH FACTORS

Though a different cell source may ultimately be used for meniscal engineering, an understanding of the behavior of fibrochondrocytes to growth factors is advantageous as it provides the framework for culturing meniscus-like cells. Meniscal fibrochondrocytes respond in different ways to different growth factors, as outlined in Table 3.6. It has been shown in multiple studies that monolayer cultures of fibrochondrocytes exposed to transforming growth factor beta 1 (TGF-β1) exhibit enhanced proteoglycan synthesis and cell proliferation [18, 43, 169]. Fibroblast growth factor (FGF), hepatocyte growth factor (HGF), platelet-derived growth factor (PDGF-AB), bone morphogenetic protein 2 (BMP-2), and human platelet lysate (Human PL) also increase cell proliferation, but at a high enough concentration FGF decreases GAG synthesis [40, 154, 170]. Cell migration of all fibrochondrocytes is stimulated by PDGF-AB and HGF, and slightly by epidermal growth factor (EGF), but only outer meniscus cells migrate when exposed to interleukin-1 (IL-1) [154]. All of these growth factors have some effect on meniscal cells and therefore can be used along with scaffolds to enhance matrix synthesis, cell proliferation, and cell penetration into tissue engineered constructs.

Several of these growth factors have already been incorporated into the culture medium for engineered constructs, and their effects investigated. One growth factor, TGF-β1, has proved especially effective for enhancing construct properties in various engineering modalities. Using a scaffoldless approach in which cells are cultured in a non-adherent well, application of TGF-β1 has been shown to increase tensile properties of engineered fibrochondrocytic constructs up to 3 MPa [112]. TGF-β1 has also been shown to increase matrix production of fibrochondrocytes seeded on poly-L-lactide (PLLA) constructs, outperforming PDGF-AB, FGF, and insulin-like growth factor 1 (IGF-1) [155]. Bone marrow mesenchymal stem cells seeded on collagen matrices are also stimulated

Table 3.6: Effects of growth factors on meniscus cells

Type	Source	Effect	References
TGF-β1	Sheep	Increases proteoglycan synthesis	18, 43, 130, 169
	Human	Increases proteoglycan synthesis	
	Rabbit	Increases [^{35}S]-sulfate [^{3}H]-proline uptake,	
	Rabbit	15-fold increase in collagen production and	
		8-fold increase in GAG production	
BMP-2	Cow	Stimulates some cell migration in red-white region, also stimulates proliferation	154
IL-1	Cow	Stimulates migration of outer meniscus cells	154
PDGF-AB	Cow	Stimulates cell migration, proliferation	154
IGF-I	Cow	Stimulates some cell migration in red-white region	154
EGF	Cow	Stimulates half of inner and outer cells to migrate	154
HGF	Cow	Stimulates cell migration, proliferation	154
FGF	Rabbit	Increases proliferation	40
Human PL	Rabbit	Increases proliferation	40

by TGF-β1 to make collagen type II and glycosaminoglycans, important proteins for reconstruction of the inner one-third of the meniscus [171].

3.2.4   SYNTHETIC SCAFFOLDS

In scaffold-based approaches, scaffold material can have similar effects on cell synthetic profiles as growth factors. Scaffolds that have been used for meniscal engineering are either synthetic, such as PLLA, or natural, such as decellularized tissue or scaffolds made from various matrix proteins (see Table 3.7). Synthetic scaffold materials must be biocompatible, but have the innate advantage over natural scaffolds of low batch variability, which is important for reproducibility. For meniscus engineering polyglycolic acid (PGA) and PLLA have been used previously. These two scaffold materials are biocompatible and degradable, and their degradation products are non-toxic, though acidic. Seeding fibrochondrocytes on PGA scaffolds has the effect of increasing cell proliferation, sulfated GAG production, and collagen synthesis over an agarose scaffold control [141]. PLLA has also been proven biocompatible with fibrochondrocytes, though the effect of PLLA alone on fibrochondrocyte processes has not been investigated [155].

Table 3.7: Scaffolds and scaffold-free methods for *in vitro* meniscus engineering

Type	Details	Result	Refs.
Synthetic			
PLLA	2.5×10^7 rabbit fibrochondrocytes/cm^3, 9 days, addition of growth factors	Supports cell survival, attachment	155
PGA	23×10^6 50:50 bovine fibrochondrocytes and chondrocytes, meniscus-shaped, 8 wks	Irregularly shaped, presence of GAGs and collagen, random orientation of collagen, stiffness 16 ± 5 kPa, fibers degraded by 8 wks	103
PGA	5×10^7 rabbit fibrochondrocytes/cm^3, 7 wks	Increases sulfated GAGs, cellularity	141
Natural			
Collagen type II–GAG	18×10^6 calf or dog fibrochondrocytes/cm^3, 3 wks	<10% contraction, 2-fold increase in DNA content, GAG and type I collagen synthesis	42
Collagen type I–GAG	18×10^6 calf or dog fibrochondrocytes/cm^3, 3 wks	50% contraction, cells confined to margins of scaffold, produced GAG and collagen	42
Hyaluronic acid	3.9×10^7 bovine or human fibrochondrocytes/cm^3, 4 wks, mixed or rotating flask	Bi-zonal tissue created with meniscus-like collagen organization, matrix deposition, and mechanical behavior	172, 173
Agarose	5×10^7 rabbit fibrochondrocytes/cm^3, 7 wks	Some initial cell death, rounded cell morphology, some GAG and collagen production	141
Decellularized meniscus	10^5 sheep fibrochondrocytes/ml, 4 wks	Scaffold nontoxic, higher stiffness (17%) and compression (26%) than native	174

Continues

Table 3.7: *Cont.* Scaffolds and scaffold-free methods for *in vitro* meniscus engineering

Type	Details	Result	Refs.
Scaffold-free			
Self-assembly	50:50 co-culture of bovine fibrochondrocytes and chondrocytes, meniscus-shaped agarose well, 8 wks	Stiffness anisotropic, circumferential modulus 226 ± 76 kPa, radial modulus 67 ± 32 kPa, meniscus-like collagen orientation, presence of GAGs and collagen types I and II	103
Self-assembly	50:50 co-culture of bovine fibrochondrocytes and chondrocytes, agarose well, 4 wks	Collagen I and II present, GAGs, constructs largely uncontracted	113
Self-assembly	Bovine fibrochondrocytes, agarose well, 4 wks	Collagen I and GAGs present, constructs contracted significantly	113

3.2.5   NATURAL SCAFFOLDS

Natural scaffolds have also garnered success in meniscal engineering (see Table 3.7). Collagen meshes have been shown to allow cell proliferation and collagen and GAG production, and to allow for growth factors to enhance these properties [46, 171]. The type of collagen used for the scaffold also has an effect on cell behavior. Collagen type II-GAG meshes fared better than type I-GAG meshes when seeded with fibrochondrocytes, reducing contraction of the constructs and increasing collagen and GAG production [42]. Matrices made of hyaluronic acid (HA) are also biocompatible, as they are made of a major component of cartilage, and when cultured in a mixed flask can create bi-zonal tissue reminiscent of the non-uniform properties of the meniscus [172, 173]. Agarose is another type of natural scaffold that has been used extensively in cartilage tissue engineering. This scaffold material can encapsulate fibrochondrocytes as it crosslinks and can be made into any shape. This material has not been shown to be as effective at stimulating matrix production or imparting mechanical integrity as other scaffold materials, however, and therefore is not as popular for meniscal engineering [141]. The last type of natural scaffold material used in meniscal engineering is decellularized meniscal tissue. This material is advantageous because it contains matrix that is already organized and can best simulate the natural microenvironment for meniscal cells. While decellularization of meniscal tissue has been achieved, mechanical integrity of the matrix has been maintained, and cytocompatibility has been demonstrated, work has yet to be done to demonstrate its use in tissue engineering [174].

3.2.6   SCAFFOLD-FREE APPROACHES

An emerging technique in meniscal tissue engineering is to use a scaffold-free method to grow meniscal tissue. Eliminating scaffold material from the tissue engineering approach eliminates variables such as the degradation profile, acidic degradation products from polymers such as polylactides/polyglycolides, and biocompatibility of the scaffold. Recently, a self-assembling method has been devised for scaffoldless tissue engineering in which cells are seeded into an agarose mold which inhibits cell attachment [104]. Using the self-assembling method, a 50:50 co-culture of bovine fibrochondrocytes and chondrocytes were formed into constructs which were cultured for 4 weeks and were found to contain both collagen types I and II as well as proteoglycans [113].

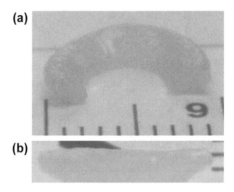

Figure 3.3: **Self-assembled meniscus-shaped construct.** The length (a) and height (b) in millimeters of a self-assembled construct after 4 weeks in culture is shown. The construct correctly mimics the wedge-shaped, semi-circular geometry of the native meniscus.

This method has also been employed to create meniscus-shaped constructs using this same co-culture of bovine meniscal fibrochondrocytes and chondrocytes, showing that these constructs were 200–400% stiffer after 4 weeks than cell-seeded PGA scaffolds [103]. This scaffold-free method therefore seems promising for recreating meniscus biochemistry and geometry as well as improving upon mechanical properties of constructs formed using synthetic scaffolds (see Table 3.7).

CONCEPTS

In vitro tissue engineering of the knee meniscus has yet to produce constructs that match native meniscal properties. A variety of cell types are under investigation for meniscal engineering, prompted by the overall scarcity of fibrochondrocytes. Progenitor cells, either adult or embryonic, hold much promise as they have shown chondrogenic capacity when exposed to various growth factor treatments. Much work has been done to identify biochemical factors and scaffold materials to which fibrochondrocytes are sensitive, and various studies have used them in combination to maximize construct biochemistry and biomechanics. Much is still unknown as to the effect of other combinations of cells and scaffold materials as well as the addition of mechanical stimulation to many of the current approaches. Some promising results have been obtained using co-cultures of chondrocytes and fibrochondrocytes in a scaffold-free approach (self-assembly process), which mimic the native meniscus in terms of biochemical makeup and geometrical properties as well as improve upon the mechanical properties when compared to constructs that use synthetic scaffolds.

3.3   *IN VIVO* TISSUE ENGINEERING

3.3.1   INTRODUCTION

As a next step to *in vitro* studies or to demonstrate feasibility of a tissue engineering technology, *in vivo* work with the knee meniscus provides practical measures of construct properties. There have been many different attempts to repair damaged meniscal tissue by implanting scaffolds with or without cells into the body, using the joint itself as a bioreactor to stimulate tissue formation.

3.3.2   ANIMAL MODELS

An ideal model for testing engineered meniscus products would replicate the environment found in the human knee joint. Though this has not yet been achieved, various animal models are used as approximations, providing a first understanding of how engineered meniscus tissue will fare when placed inside an organism. The most popular animal model for testing engineered meniscus technologies is the New Zealand white rabbit [175]–[181]. Well-characterized and relatively inexpensive, the rabbit model is large enough for surgical procedures and small enough to raise easily. Though it has been shown that the mechanical properties of the rabbit meniscus are not as similar to the human meniscus as those of baboons, dogs, and pigs, the rabbit still remains the model of choice for most studies [37]. Larger animal models such as dogs, sheep, pigs, and goats have also been used to show the biocompatibility and feasibility of tissue engineered meniscus constructs [182]–[188]. Showing good performance in these models may more closely approach expected outcomes in a human, as they are more comparable in size and the meniscus in these animals has similar biomechanical properties to human tissue. Additionally, testing in these types of animals is often required by the Food and Drug Administration (FDA) prior to embarking on clinical trials. Table 3.8 lists the advantages and disadvantages of the various animal models used in meniscus tissue engineering. It is important to note that in addition to the type of animal used, the age of the animal has a large influence on tissue plasticity. Researchers must take the age of the animal model into account when evaluating *in vivo* success rates.

3.3.3   FIBRIN

The earliest attempts to tissue engineer cartilage *in vivo* involved using fibrin or fibrin clots as a scaffold. Fibrin, a natural material formed in the body following injury, can be processed into a gel and in combination with endothelial cell growth factor was shown to increase healing of defects created in the dog meniscus [187]. Following this success, fibrochondrocytes were encapsulated in

Animal model	Advantages	Disadvantages	References using
Rabbit	Commonly used for meniscus studies, cost, feasibility, relatively easily operable, small	Small, meniscal properties differ from humans	175-181
Dog	Large, meniscus Poisson's ratio and permeability similar to humans, easily operable	Cost, feasibility	186, 187
Sheep	Large, easily operable	Cost, feasibility	184, 185
Pig	Large, meniscus Poisson's ratio similar to humans, easily operable	Cost, feasibility	188
Goat	Large, easily operable	Cost, feasibility	182, 183
Human	Most clinically relevant	Regulatory approval	189, 190

Table 3.8: Animal models used in meniscus tissue engineering

fibrin gels and implanted into rabbit meniscal defects. After two months, the scaffolds showed signs of cell proliferation and sulfated GAG production [181]. Fibrin gel has shown better success than fibrin clots even with the incorporation of mesenchymal stem cells. Fibrin clots that were implanted into defect sites did not induce much healing when implanted into avascular defect sites of the meniscus [183]. Although some success has been observed with these fibrin scaffolds, full defect healing was not observed and no mechanical testing was performed on the repair tissue.

3.3.4   SYNTHETIC SCAFFOLDS

In the mid-1990's, cell-seeded synthetic scaffolds were investigated *in vivo* as possible replacement therapies for meniscal defects (see Table 3.9). PGA scaffolds were seeded with meniscal fibrochondrocytes and implanted subcutaneously in nude mice for 16 weeks, showing organization with time of a fibrous matrix containing sulfated GAGs [191]. These scaffolds were also seeded with fibrochondrocytes transfected with a gene to encourage vascularization, and implanted subcutaneously. In this experiment the gene transfection successfully encouraged vascularization of the construct [192]. Recently, a PGA scaffold was seeded with fibrochondrocytes and implanted into a rabbit knee that had undergone total meniscectomy. Following 10 weeks *in vivo*, fibrocartilaginous tissue had formed, although the biochemical and mechanical characteristics of the neotissue were still a fraction of native tissue [176]. Another synthetic scaffold type comprised of a PLLA and poly(p-dioxanone) PPD blend was also used recently as a meniscal prosthesis to stimulate meniscal regeneration [177]. After 14 weeks there was some ingrowth of tissue and extensive degradation of the scaffold [177]. Additionally the cartilage underlying the scaffold was intact, while in control specimens the cartilage

Table 3.9: Scaffolds for *in vivo* meniscus engineering

Type	Details	Result	Ref.
Synthetic			
PGA	2.5×10^7 bovine fibrochondrocytes/mL alginate, subcutaneous nude mouse, 16 wks	Fibrous matrix formation, presence of GAGs and collagen	191
PGA	2×10^7 bovine fibrochondrocytes/cm^3, subcutaneous nude mouse, 8 wks	Fibrous matrix formation, firm consistency, collagen and GAG presence	192
PGA	2×10^6 rabbit fibrochondrocytes/scaffold, meniscus-shaped scaffold, replaced meniscus in rabbit knee, 10 wks	Fibrous tissue formation, up to 40% collagen/dry weight, GAG and collagen type II found in inner portion of scaffold, type I collagen in outer portion	176
PLLA/PPD	acellular, filled rabbit partial meniscectomy defect, 14 wks	Fibrous matrix formation, some collagen alignment, some preservation of underlying hyaline cartilage	177
Natural			
Fibrin	Rabbit fibrochondrocytes, 8 wks	Cells able to produce sulfated GAGs, proliferate	181
Small intestine submucosa	Filled goat meniscal defect, 12 wks	Fibrous tissue formation, no organization, partially filled defect, hyaline cartilage degeneration	182
Acellular meniscus	Pig chondrocytes on surface of scaffold, filled longitudinal tear in avascular region of pig meniscus, 9 wks	Some healing of tear observed, matrix contained GAGs	188

Continues

Table 3.9: *Cont.* Scaffolds for *in vivo* meniscus engineering

Type	Details	Result	Ref.
Natural (cont.)			
Hyaluronan	Rabbit bone-marrow mesenchymal stem cells, rabbit meniscal defect, 12 wks	Some repair tissue formed, contained collagen type II	175
Collagen meniscus implant	Acellular, human meniscal defect, 5 years	Pain scores improved significantly, implant maintained structure	189
Collagen meniscus implant	Sheep fibrochondrocytes, sheep meniscal defect, 21 wks	Scaffold contraction, some fibrous repair tissue formed	184

had signs of damage [177]. Therefore, the PLLA/PPD scaffold was shown effective as a meniscal prosthetic but was not able to stimulate complete tissue repair prior to degradation of the material.

3.3.5   NATURAL SCAFFOLDS

Currently, focus has turned to the potential of natural scaffolds to repair the meniscus (see Table 3.9). Studies using decellularized tissue such as the meniscus and small intestine submucosa have been able to achieve partial regeneration of meniscal tears, but their *in vivo* success has been limited [182, 188]. More success has been achieved using collagen scaffolds from bovine Achilles tendon called collagen meniscus implants (CMI). These implants aim to act as prostheses and to stimulate tissue ingrowth. Longitudinal studies in patients have shown that 6 months post-implantation, the underlying cartilage was intact, the scaffold was infiltrated with fibrochondrocytes and that the implant contained fibrous tissue after 5 years [189, 190]. Another study in sheep used CMI in a tissue engineering strategy in which fibrochondrocytes were seeded into the scaffold and then implanted in an inner-meniscus defect [184]. The results after 3 months suggested that cell-seeded CMI performed better than non-seeded controls, although the matrix produced was more fibrous than that found in the inner meniscus, and the constructs tended to contract significantly [184]. CMI is therefore a promising technology for meniscal replacement, although more research must be done to induce inner-meniscus like tissue to form.

Another natural scaffold used currently in meniscal tissue engineering research *in vivo* is hyaluronan. This scaffold has been used to encapsulate chondro-differentiated mesenchymal stem cells and was implanted in a rabbit meniscal defect for 12 weeks [175]. The repair tissue formed by this method was more similar to native meniscus tissue than non-treated controls and good integration was observed between the scaffold and native meniscus [175].

CONCEPTS

In vivo testing of engineered materials is important for ensuring the biocompatibility of the technology as well as its feasibility to repair damaged tissue. The most popular animal model for *in vivo* testing of meniscus constructs is the New Zealand white rabbit. Other animals that have been used include sheep, dogs, pigs, and goats. Many different types of cells and scaffold materials have been investigated *in vivo* for use as meniscal replacements. Fibrin has demonstrated a limited capacity to induce sufficient repair of meniscal tissue. The addition of cells to scaffolds tends to increase overall reparative potential of a construct, but complete healing of a meniscus defect has yet to be achieved. A number of scaffold materials have been used for meniscus tissue engineering including synthetic types (PGA, PPD, PLLA) and natural types (collagen, hyaluronan). Recent focus has been on using natural scaffolds rather than synthetic ones to better achieve meniscus-like repair tissue. There are a few promising approaches currently being investigated using both types of scaffolds, but more research must be done to understand their long-term behavior in the body.

CHAPTER   4

Current Therapies and Future Directions

4.1   PRODUCTS AND CURRENT THERAPIES

4.1.1   PRODUCTS INVOLVING BIOLOGICAL MATERIALS

Currently, the only clinical product available for meniscus repair that uses biological materials is the Menaflex^{TM} collagen meniscus implant, which has been approved for use in Europe and, more recently, in the United States. Developed in the United States by ReGen Biologics (Hackensack, New Jersey), the implant is approved for use in both the medial and lateral menisci in Europe, but in the U.S. it is currently only FDA approved for use in the medial meniscus. Menaflex^{TM} is made of a collagen type I scaffold which can be sutured to the remaining meniscus following partial meniscectomy and degrades almost completely after one year. It has gone through several clinical trials and has shown an ability to allow cell infiltration and increase activity levels of patients who had a history of meniscus surgeries [193]. For patients with acute injuries who had not undergone previous surgeries on the meniscus, activity level increased the same amount with or without treatment with Menaflex^{TM} [193]. While activity levels do increase for patients with chronic meniscus problems, arthroscopic evaluation has revealed that this treatment will only partially fill a meniscus defect site (around 50–60% on average) [193]. Additionally, the repair tissue resulting from Menaflex^{TM} treatment is not extensively characterized, leaving the biomechanical benefits of the implant unknown. Therefore, though there exists a product on the market to address the need for replacement meniscus tissue, it may not be optimal in all cases and there is still a need for more alternatives.

4.1.2   OTHER CURRENT THERAPIES

Clinical therapies for meniscus repair are varied. Meniscus surgery is most often performed arthroscopically, as this method is the least invasive and leaves the smallest scars. Depending on the type of meniscus injury, surgical techniques may involve arthroscopic suturing, other fixation devices, or meniscectomy. Abrasion may also be used to smooth the remaining meniscus tissue following meniscectomy. The success of sutures or other fixation devices to encourage healing of meniscus tears depends on the location of the tear. Tears in the outer portion of the meniscus are more likely to bond when held together due to the proximity of blood vessels, whereas tears in the inner region are unlikely to repair.

Suturing is a popular repair mode for bucket-handle tears as it can hold a tear closed to encourage bonding. Suture materials in use today are usually biodegradable, eliminating the need for subsequent procedures for removal. Suturing techniques can be inside-out, outside-in, or all-inside depending on where the suture material first enters the meniscus (see Figure 4.1) [194, 195].

Suturing meniscus tears can take around 20 minutes using the inside-out technique, and around 40 minutes using the outside-in technique [196]. Both of these suturing methods have around 100% success rates for healing a peripheral tear [196]. All-inside suturing techniques are less common, but can provide the benefit of reducing arthroscopic incision size compared to the other methods [195], [197]–[199].

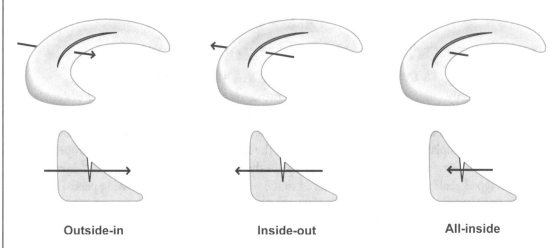

Outside-in   **Inside-out**   **All-inside**

Figure 4.1: **Meniscus suturing techniques.** There are three main modes of suturing the meniscus arthroscopically. The outside-in technique passes the suture material from the periphery of the tissue to the inner surface, while the inside-out technique passes the suture into the meniscus from within the joint space, exiting at the periphery. The all-inside technique does not affect as much of the meniscus, as the suture enters and exits only at the inner portion of the tissue. Other repair methods such as the meniscus arrow are considered all-inside techniques as they are implanted into the tissue via the inner surface.

Recently, all-inside repair techniques that do not use sutures have gained popularity. These include meniscus arrows, screws, and anchors. These all-inside fixation devices take less time to place than normal sutures, though there is debate as to whether these methods work as well as suturing [196], [200]–[207]. Amongst the all-inside repair options, the most popular is the meniscus arrow. The meniscus arrow is quick and easy to implement, uses fewer incisions than suturing techniques, and has a higher success rate than either screws or anchors. Success rates for meniscus arrow treatments typically range from 70–95% [203, 204, 208]. In contrast, success rates for meniscus screws and anchors are reported to be around 27% and 65%, respectively [196, 207]. Though there are many benefits to using meniscus arrows, they also involve some risks including migration out of the tissue, inflammation of the surrounding tissue, and/or damage to the hyaline cartilage surfaces [202, 205].

In some cases, the meniscus is so damaged that total meniscectomy is necessary. In this case, the knee is either left without a meniscus or a meniscus allograft is implanted. Allografts, though scarce, can improve the stability of the knee and increase the success rate of ACL repair [209]. In arthritic joints, a meniscus allograft has an estimated lifetime of around 4.4 years and can decrease joint pain significantly [210]. Challenges to using meniscus allografts on a large scale include difficulty in preserving, sterilizing, and attaching the tissue, as well as the limited lifetime of the material [211].

Partial meniscectomy and abrasion are used to remove torn inner meniscus tissue that causes pain or impedes joint movement. In arthroscopic partial meniscectomy, a scalpel or other tool is used to cut away the torn tissue, leaving the rest of the meniscus intact. Following partial meniscectomy, abrasion may be performed to smooth out the remaining meniscal surface, thereby discouraging tear propagation or new tear formation. This mode of therapy was shown especially effective for tears of the posterior horn of the lateral meniscus, or radial tears that appear stable [212].

CONCEPTS

Presently, the only product available for meniscus repair that uses biological materials is the Menaflex^{TM} collagen meniscus implant by ReGen Biologics. This product is a collagen type I sponge that can be sutured to the remaining meniscus after a partial meniscectomy. Menaflex^{TM} is biocompatible, providing a scaffold for cell infiltration and new tissue formation, and has increased activity levels in some patients. Use of this product, however, has not demonstrated full defect healing. Other current therapies that do not involve engineered biological materials include sutures and other fixation devices (arrows, screws, anchors), meniscectomy, abrasion, and allografting. Suturing is most successful for vertical-longitudinal (bucket-handle) tears in the outer region of the meniscus and can be done three ways: outside-in, inside-out, or all-inside. The all-inside technique has become popular with the introduction of meniscus arrows, which are easy and quick to implant as compared to suturing techniques. There is still controversy as to whether arrows work as well as sutures, and indeed the reported success rates for arrows are lower than for sutures. For inner region tears that are symptomatic, partial meniscectomy is often performed. Following partial meniscectomy abrasion of the remaining tissue can be performed to smooth out the tissue surface and discourage new tear formation. If the meniscus is thoroughly damaged, total meniscectomy may be performed and a meniscus allograft can be implanted. Allografts tend to work well for most patients, but they are generally scarce and conditioning them for implantation may compromise tissue integrity.

4.2   DESIGN STANDARDS FOR TISSUE ENGINEERING THE MENISCUS

4.2.1   DETERMINING DESIGN STANDARDS

Meniscus tissue engineering has made great advances in recent years, but has yet to produce a therapy that can fully restore function to a damaged meniscus. In the normal meniscus, a complex set of characteristics work together to form a tissue that is able to bear load, transfer stresses, absorb shock, and stabilize the knee joint. Ideally, a meniscal replacement would have the same properties as the native meniscus, but in practice it is difficult to get all of them to coexist using current tissue engineering modalities. Therefore, tissue engineers must carefully consider which meniscus properties they will reproduce such that the resulting tissue is able to perform well in the body. Based on what is known about the role of the meniscus, there are certain characteristics that a meniscal replacement must have in order to restore function. Primarily, a meniscus replacement must be biocompatible and have specific geometrical and biomechanical attributes to operate well in the knee joint. Of secondary concern are the biochemical characteristics, vascularization, enervation, and cellularity of the replacement. Prioritizing meniscal attributes as such is advantageous to the tissue engineer as it allows for the design of a tissue that is maximally functional, a main concern for meniscus replacement options.

4.2.2   PRIMARY STANDARDS

The main design standards important for tissue engineering a functional meniscus are proper geometry, biomechanics, and biocompatibility. The geometry of the meniscus has a central role in allowing for smooth joint movement and stabilization. Being wedge-shaped and semi-circular, the meniscus increases the congruence of the femur and tibia. The shape and size of tissue engineered constructs should therefore match that of the native meniscus to best restore meniscal function.

Biomechanical properties of the meniscus are also primary contributors to normal meniscus function. The anisotropic material behavior of the meniscus in tension, compression, and shear allows for dynamic load bearing and distribution and is imparted to the tissue through the organization of matrix molecules. Engineered meniscus tissue must also withstand these loading patterns, either by organized matrix deposition or other means. Specifically, circumferential tensile properties for the engineered tissue must be higher than radial properties, and compressive properties must be higher in the inner portion than in the outer portion. Having mechanical properties regionally similar to native values will ensure that the construct is not destroyed upon implantation. This is especially important if the engineered tissue is used to fill a defect site, in which case mechanical similarity will ensure a normal distribution of load. Creating a biomechanically robust tissue is also advantageous for surgical implantation. Fixation methods involving screws and sutures place great amounts of local stress on the tissue. As it is known that the meniscus relies heavily on tibial horn attachments for anchoring within the joint, it is imperative that a replacement tissue is able to withstand these fixation stresses in addition to loading stresses.

The safety of an implantable meniscal replacement is not only linked to its function within the joint, but also to its immune response. Biocompatibility is, of course, a primary concern for tissue engineering the meniscus as this attribute can increase safety, which is highly important for eventual clinical approval. Increasing the biocompatibility of the engineered tissue by using natural scaffolds or scaffolds with non-toxic degradation products, or using autologous cells can reduce the likelihood of an immune reaction in the body.

4.2.3   SECONDARY STANDARDS

Creating a tissue that mimics the geometrical, biomechanical, and biocompatibility of the native meniscus may be sufficient to restore function, but there are some secondary characteristics that could increase its longevity and overall compatibility within the body. These include biochemical makeup, vascularization, and cellularity. Once the primary functional characteristics are achieved in a meniscal replacement, these secondary characteristics should be examined.

There are many different chemical components that make up meniscus tissue, yet all of their functions are not known. Given their relative abundance in the tissue, however, collagens and proteoglycans seem to be the main biochemical contributors. Their distributions and orientations within the meniscus contribute greatly to the regional variation in mechanical properties, highlighting the need for tissue with complex regional makeup for replacement. Specifically, collagen is primarily oriented circumferentially, and proteoglycans are most abundant in the inner portion of the meniscus. Collagen type I is the primary type of collagen in the outer portion, while collagen type II is more abundant in the inner portion. Mimicking the regionally varying relative abundance of these two components of the meniscus, as well as their structural organization, may lead to appropriate mechanical properties. It should be noted, however, that other molecules such as those used for adhesion may be important for tissue organization and overall function. The functions of many of these minor biochemical components of the meniscus are not well understood, but they could play an important role in creating a functional construct.

Because not all of the functions of a meniscal blood supply are known, the importance of vascularization for a tissue engineered meniscus is difficult to determine. In the normal adult meniscus, the blood supply is confined to the peripheral two-thirds and helps in tissue remodeling following injury. It is likely also responsible for proper delivery of nutrients and removal of waste products. Depending on the cellularity and permeability of the engineered tissue, vascularization could play an important role in allowing the tissue to thrive indefinitely.

The cellular aspects of an engineered meniscus are also of secondary concern. As many as four different types of cells have been identified in the meniscus, though their functions are not well understood. The morphologies of these cells range from rounded in the inner portion of the meniscus to having many cell processes in the outer portion. While in general cells are important for the creation or remodeling of tissue, little is known about the extent of remodeling that takes place in the adult meniscus. For engineered tissue, cells are important for creating specific types of matrix molecules that can aid in developing a mechanically robust construct. It is unclear, however, whether having the exact types and morphologies of cells found in the native meniscus is necessary for maintaining the overall function of an engineered meniscus.

CONCEPTS

In designing an engineered meniscal replacement, much thought must be put into which characteristics are most important to the overall function of the construct. For the meniscus, a biomechanically active tissue, primary design standards focus on recreating functional aspects. These include mimicking meniscus geometry, biomechanics, and biocompatibility. Of secondary concern are the specific biochemical makeup of the construct, its vascularity, and the types of cells that it contains. By focusing primarily on the functional aspects, a robust and clinically suitable tissue may be achieved. As more is discovered about the specific functions of meniscus matrix molecules, vascular supply, and cells, these secondary criteria may become useful to extend construct longevity and compatibility in the body.

4.3 ASSESSMENTS FOR TISSUE ENGINEERED CONSTRUCTS

4.3.1 NEED FOR FUNCTIONAL ASSESSMENT

Whether or not engineered meniscus tissue will perform well *in vivo* may depend on many different factors, including biochemical and biomechanical properties of the engineered neotissue. Not only do laboratories engaged in tissue engineering need standards to evaluate the functionality of their work, but regulatory agencies must likewise determine the type of assessments required for approval of such products.

4.3.2 FUNCTIONALITY INDEX

Common methods to functionally evaluate engineered meniscus tissue are tensile tests, compression tests, shear tests, and biochemical assays. Ideally, all of these measurements would be taken, and the collective set of properties would be compared to native tissue. Typically, however, only one measurement is compared to the gold standard at a time, giving a limited view of the overall functionality of the engineered construct. Determining this overall functionality is not trivial, as some construct properties will inevitably be closer to native values than others. Using a functionality index (FI) is one way to assess the overall similarity of an engineered tissue to known native values. This method uses a weighted average taken of the similarity fractions of the engineered tissue's properties compared to native values. The FI has been used previously to compare engineered hyaline cartilage to native hyaline cartilage [213].

$$FI = \frac{1}{4} \left(\begin{array}{c} \left(1 - \frac{(G_{nat}-G_{ec})}{G_{nat}}\right) + \left(1 - \frac{(C_{nat}-C_{ec})}{C_{nat}}\right) + \frac{1}{2}\left(1 - \frac{(E_{nat}^T-E_{ec}^T)}{E_{nat}^T}\right) + \\ \frac{1}{2}\left(1 - \frac{(E_{nat}^C-E_{ec}^C)}{E_{nat}^C}\right) + \frac{1}{2}\left(1 - \frac{(S_{nat}^T-S_{ec}^T)}{S_{nat}^T}\right) + \frac{1}{2}\left(1 - \frac{(S_{nat}^C-S_{ec}^C)}{S_{nat}^C}\right) \end{array} \right) \tag{4.1}$$

In the functionality index above (Equation 4.1), G and C represent the total GAG and total collagen per wet weight, E represents stiffness and S represents strength. Superscripts of T and C indicate tensile and compressive properties, respectively, and the subscripts '*nat*' and '*ec*' indicate natural or engineered construct properties, respectively. According to this equation, equal weight is given to biochemical and mechanical properties. This could be adjusted, however, if one property demonstrates a dominant role in the overall functionality of the tissue. As the right-hand side of the equation approaches a value of 1, the engineered tissue approaches the properties of native tissue. A value close to zero indicates that the engineered tissue has much lower properties than native tissue, and a value higher than 1 indicates that some or all of the engineered tissue's properties are higher than native values. The functionality index can therefore provide an extra measure of engineered tissue's similarity to native tissue by considering all measured properties simultaneously.

4.3.3   VARIABLE CONSIDERATIONS

There are many different considerations to make when writing a functionality index equation. Weights must be assigned to each parameter based on its overall importance to tissue functionality. Whether to assign mature or immature tissue properties to the native tissue variables in the equation must also be determined. Once the equation is formulated, a level must be set to determine satisfactory or unsatisfactory similarity to the comparison tissue.

In the above example, equal weight is applied to biochemical and mechanical properties. For the meniscus, as it is known that mechanical integrity plays a large functional role, more weight might be shifted toward mechanical performance. Other parameters that take into account the circumferential versus radial stiffness and strength could also be included in the equation to assess appropriate functional behavior.

As engineered meniscal tissue is likely to be implanted into adult patients, using adult tissue properties as a comparison in the FI equation could provide valuable information. During the process of tissue development, however, engineered meniscus constructs may exhibit immature properties. Immature native tissue properties could therefore be used in the FI during this process to determine if engineered tissue mimics native tissue in early developmental stages. While it is unclear whether native properties should be attained prior to implantation or if a construct can mature *in vivo*, the FI is a useful tool to determine similarity between engineered and native tissues and may eventually be used to identify the readiness of a construct for implantation.

A highly important consideration for the tissue engineer is how similar an engineered tissue must be to native tissue in order to be useful in the body. The functionality index is suited to quantify this similarity, but it remains to be determined what FI threshold must be reached to indicate acceptable properties. More research must be done to determine suitable characteristics for implantation, but once this threshold is determined, engineers will have a tangible success criterion that takes into account many different aspects of an engineered tissue.

CONCEPTS

Engineers use various tests to functionally assess their tissue engineered constructs. The properties that are measured are usually considered alone, allowing for only a limited understanding of tissue functionality. The functionality index (FI) can be used to consider all measured properties of an engineered construct compared to native tissue. This allows for an overall assessment of tissue functionality. The FI is calculated by taking the average of similarity fractions of engineered tissue properties compared to native tissue properties. An unlimited number of properties can be included in this equation, and different weights can be assigned to each property based on its importance to tissue performance. When writing a functionality index equation for engineered meniscus tissue, it may be necessary to weight certain mechanical properties higher than biochemical properties, given its highly biomechanical role in the body. Also of importance is determining what type of tissue (mature or immature) to use for comparison, as well as identifying an FI value that indicates an acceptably functional tissue.

Bibliography

[1] Ghadially, F.N., I. Thomas, N. Yong, and J.M. Lalonde, *Ultrastructure of rabbit semilunar cartilages.* J Anat, 1978. **125**(Pt 3): p. 499-517.

[2] Tham, S.C., I.Y. Tsou, and T.S. Chee, *Knee and ankle ligaments: magnetic resonance imaging findings of normal anatomy and at injury.* Ann Acad Med Singapore, 2008. **37**(4): p. 324-6.

[3] Brindle, T., J. Nyland, and D.L. Johnson, *The Meniscus: Review of Basic Principles With Application to Surgery and Rehabilitation.* J Athl Train, 2001. **36**(2): p. 160-169.

[4] Greis, P.E., D.D. Bardana, M.C. Holmstrom, and R.T. Burks, *Meniscal injury: I. Basic science and evaluation.* J Am Acad Orthop Surg, 2002. **10**(3): p. 168-76.

[5] Clark, C.R. and J.A. Ogden, *Development of the menisci of the human knee joint. Morphological changes and their potential role in childhood meniscal injury.* J Bone Joint Surg Am, 1983. **65**(4): p. 538-47.

[6] Pufe, T., W.J. Petersen, N. Miosge, M.B. Goldring, R. Mentlein, D.J. Varoga, and B.N. Tillmann, *Endostatin/collagen XVIII–an inhibitor of angiogenesis–is expressed in cartilage and fibrocartilage.* Matrix Biol, 2004. **23**(5): p. 267-76. DOI: 10.1016/j.matbio.2004.06.003

[7] O'Reilly, M.S., T. Boehm, Y. Shing, N. Fukai, G. Vasios, W.S. Lane, E. Flynn, J.R. Birkhead, B.R. Olsen, and J. Folkman, *Endostatin: an endogenous inhibitor of angiogenesis and tumor growth.* Cell, 1997. **88**(2): p. 277-85. DOI: 10.1016/S0092-8674(00)81848-6

[8] McAlinden, A., J. Dudhia, M.C. Bolton, P. Lorenzo, D. Heinegard, and M.T. Bayliss, *Age-related changes in the synthesis and mRNA expression of decorin and aggrecan in human meniscus and articular cartilage.* Osteoarthritis Cartilage, 2001. **9**(1): p. 33-41. DOI: 10.1053/joca.2000.0347

[9] McDevitt, C.A. and R.J. Webber, *The ultrastructure and biochemistry of meniscal cartilage.* Clin Orthop Relat Res, 1990. (252): p. 8-18. DOI: 10.1097/00003086-199003000-00003

[10] Ghosh, P. and T.K. Taylor, *The knee joint meniscus. A fibrocartilage of some distinction.* Clin Orthop Relat Res, 1987. (224): p. 52-63. DOI: 10.1097/00003086-198711000-00008

[11] McNicol, D. and P.J. Roughley, *Extraction and characterization of proteoglycan from human meniscus.* Biochem J, 1980. **185**(3): p. 705-13.

[12] Sweigart, M.A. and K.A. Athanasiou, *Toward tissue engineering of the knee meniscus.* Tissue Eng, 2001. **7**(2): p. 111-29. DOI: 10.1089/107632701300062697

[13] Wojtys, E.M. and D.B. Chan, *Meniscus structure and function.* Instr Course Lect, 2005. **54**: p. 323-30.

[14] Cheung, H.S., *Distribution of type I, II, III and V in the pepsin solubilized collagens in bovine menisci.* Connect Tissue Res, 1987. **16**(4): p. 343-56. DOI: 10.3109/03008208709005619

[15] Hopker, W.W., G. Angres, K. Klingel, D. Komitowski, and E. Schuchardt, *Changes of the elastin compartment in the human meniscus.* Virchows Arch A Pathol Anat Histopathol, 1986. **408**(6): p. 575-92.

[16] Herwig, J., E. Egner, and E. Buddecke, *Chemical changes of human knee joint menisci in various stages of degeneration.* Ann Rheum Dis, 1984. **43**(4): p. 635-40. DOI: 10.1136/ard.43.4.635

[17] Scott, P.G., T. Nakano, and C.M. Dodd, *Isolation and characterization of small proteoglycans from different zones of the porcine knee meniscus.* Biochim Biophys Acta, 1997. **1336**(2): p. 254-62.

[18] Tanaka, T., K. Fujii, and Y. Kumagae, *Comparison of biochemical characteristics of cultured fibrochondrocytes isolated from the inner and outer regions of human meniscus.* Knee Surg Sports Traumatol Arthrosc, 1999. **7**(2): p. 75-80. DOI: 10.1007/s001670050125

[19] McDermott, I.D., F. Sharifi, A.M. Bull, C.M. Gupte, R.W. Thomas, and A.A. Amis, *An anatomical study of meniscal allograft sizing.* Knee Surg Sports Traumatol Arthrosc, 2004. **12**(2): p. 130-5. DOI: 10.1007/s00167-003-0366-7

[20] Shaffer, B., S. Kennedy, J. Klimkiewicz, and L. Yao, *Preoperative sizing of meniscal allografts in meniscus transplantation.* Am J Sports Med, 2000. **28**(4): p. 524-33.

[21] Paul, J.P., *Force actions transmitted by joints in the human body.* Proc R Soc Lond B Biol Sci, 1976. **192**(1107): p. 163-72.

[22] Shrive, N.G., J.J. O'Connor, and J.W. Goodfellow, *Load-bearing in the knee joint.* Clin Orthop Relat Res, 1978. (131): p. 279-87. DOI: 10.1097/00003086-197803000-00046

[23] Walker, P.S. and J.V. Hajek, *The load-bearing area in the knee joint.* J Biomech, 1972. **5**(6): p. 581-9. DOI: 10.1016/0021-9290(72)90030-9

[24] Walker, P.S. and M.J. Erkman, *The role of the menisci in force transmission across the knee.* Clin Orthop Relat Res, 1975. (109): p. 184-92. DOI: 10.1097/00003086-197506000-00027

[25] Kurosawa, H., T. Fukubayashi, and H. Nakajima, *Load-bearing mode of the knee joint: physical behavior of the knee joint with or without menisci.* Clin Orthop Relat Res, 1980. (149): p. 283-90. DOI: 10.1097/00003086-198006000-00039

[26] Favenesi, J.A., J.C. Shaffer, and V.C. Mow, *Biphasic mechanical properties of knee meniscus.* Trans. 29th Annual Orthopaedic Research Society, 1983.

[27] Proctor, C.S., M.B. Schmidt, R.R. Whipple, M.A. Kelly, and V.C. Mow, *Material properties of the normal medial bovine meniscus.* J Orthop Res, 1989. **7**(6): p. 771-82. DOI: 10.1002/jor.1100070602

[28] Ghadially, F.N., J.M. Lalonde, and J.H. Wedge, *Ultrastructure of normal and torn menisci of the human knee joint.* J Anat, 1983. **136**(Pt 4): p. 773-91.

[29] Gabrion, A., P. Aimedieu, Z. Laya, E. Havet, P. Mertl, R. Grebe, and M. Laude, *Relationship between ultrastructure and biomechanical properties of the knee meniscus.* Surg Radiol Anat, 2005. **27**(6): p. 507-10. DOI: 10.1007/s00276-005-0031-6

[30] Skaggs, D.L., W.H. Warden, and V.C. Mow, *Radial tie fibers influence the tensile properties of the bovine medial meniscus.* J Orthop Res, 1994. **12**(2): p. 176-85. DOI: 10.1002/jor.1100120205

[31] Aspden, R.M., Y.E. Yarker, and D.W. Hukins, *Collagen orientations in the meniscus of the knee joint.* J Anat, 1985. **140**(Pt 3): p. 371-80.

[32] Fithian, D.C., M.A. Kelly, and V.C. Mow, *Material properties and structure-function relationships in the menisci.* Clin Orthop Relat Res, 1990. (252): p. 19-31.

[33] McDermott, I.D., S.D. Masouros, and A.A. Amis, *Biomechanics of the menisci of the knee.* Current Orthopaedics, 2008. **22**: p. 193-201. DOI: 10.1016/j.cuor.2008.04.005

[34] Zhu, W., K.Y. Chern, and V.C. Mow, *Anisotropic viscoelastic shear properties of bovine meniscus.* Clin Orthop Relat Res, 1994 (306): p. 34-45.

[35] Joshi, M.D., J.K. Suh, T. Marui, and S.L. Woo, *Interspecies variation of compressive biomechanical properties of the meniscus.* J Biomed Mater Res, 1995. **29**(7): p. 823-8. DOI: 10.1002/jbm.820290706

[36] Leslie, B.W., D.L. Gardner, J.A. McGeough, and R.S. Moran, *Anisotropic response of the human knee joint meniscus to unconfined compression.* Proc Inst Mech Eng [H], 2000. **214**(6): p. 631-5.

[37] Sweigart, M.A., C.F. Zhu, D.M. Burt, P.D. DeHoll, C.M. Agrawal, T.O. Clanton, and K.A. Athanasiou, *Intraspecies and interspecies comparison of the compressive properties of the medial meniscus.* Ann Biomed Eng, 2004. **32**(11): p. 1569-79. DOI: 10.1114/B:ABME.0000049040.70767.5c

[38] Hellio Le Graverand, M.P., Y. Ou, T. Schield-Yee, L. Barclay, D. Hart, T. Natsume, and J.B. Rattner, *The cells of the rabbit meniscus: their arrangement, interrelationship, morphological variations and cytoarchitecture.* J Anat, 2001. **198**(Pt 5): p. 525-35. DOI: 10.1046/j.1469-7580.2000.19850525.x

[39] Nakata, K., K. Shino, M. Hamada, T. Mae, T. Miyama, H. Shinjo, S. Horibe, K. Tada, T. Ochi, and H. Yoshikawa, *Human meniscus cell: characterization of the primary culture and use for tissue engineering.* Clin Orthop Relat Res, 2001. (391 Suppl): p. S208-18. DOI: 10.1097/00003086-200110001-00020

[40] Webber, R.J., M.G. Harris, and A.J. Hough, Jr., *Cell culture of rabbit meniscal fibrochondrocytes: proliferative and synthetic response to growth factors and ascorbate.* J Orthop Res, 1985. **3**(1): p. 36-42. DOI: 10.1002/jor.1100030104

[41] Moon, M.S., J.M. Kim, and I.Y. Ok, *The normal and regenerated meniscus in rabbits. Morphologic and histologic studies.* Clin Orthop Relat Res, 1984. (182): p. 264-9. DOI: 10.1097/00003086-198401000-00035

[42] Mueller, S.M., S. Shortkroff, T.O. Schneider, H.A. Breinan, I.V. Yannas, and M. Spector, *Meniscus cells seeded in type I and type II collagen-GAG matrices in vitro.* Biomaterials, 1999. **20**(8): p. 701-9. DOI: 10.1016/S0142-9612(98)00189-6

[43] Pangborn, C.A. and K.A. Athanasiou, *Effects of growth factors on meniscal fibrochondrocytes.* Tissue Eng, 2005. **11**(7-8): p. 1141-8. DOI: 10.1089/ten.2005.11.1141

[44] Miller, R.R. and P.A. Rydell, *Primary culture of microvascular endothelial cells from canine meniscus.* J Orthop Res, 1993. **11**(6): p. 907-11. DOI: 10.1002/jor.1100110618

[45] Melrose, J., S. Smith, M. Cake, R. Read, and J. Whitelock, *Comparative spatial and temporal localisation of perlecan, aggrecan and type I, II and IV collagen in the ovine meniscus: an ageing study.* Histochem Cell Biol, 2005. **124**(3-4): p. 225-35. DOI: 10.1007/s00418-005-0005-0

[46] Gruber, H.E., D. Mauerhan, Y. Chow, J.A. Ingram, H.J. Norton, E.N. Hanley, Jr., and Y. Sun, *Three-dimensional culture of human meniscal cells: extracellular matrix and proteoglycan production.* BMC Biotechnol, 2008. **8**: p. 54. DOI: 10.1186/1472-6750-8-54

[47] Ochi, K., Y. Daigo, T. Katagiri, A. Saito-Hisaminato, T. Tsunoda, Y. Toyama, H. Matsumoto, and Y. Nakamura, *Expression profiles of two types of human knee-joint cartilage.* J Hum Genet, 2003. **48**(4): p. 177-82. DOI: 10.1007/s10038-003-0004-8

[48] Ahluwalia, S., M. Fehm, M.M. Murray, S.D. Martin, and M. Spector, *Distribution of smooth muscle actin-containing cells in the human meniscus.* J Orthop Res, 2001. **19**(4): p. 659-64. DOI: 10.1016/S0736-0266(00)00041-3

[49] Mueller, S.M., T.O. Schneider, S. Shortkroff, H.A. Breinan, and M. Spector, *alpha-smooth muscle actin and contractile behavior of bovine meniscus cells seeded in type I and type II collagen-GAG matrices.* J Biomed Mater Res, 1999. **45**(3): p. 157-66. DOI: 10.1002/(SICI)1097-4636(19990605)45:3<157::AID-JBM1>3.3.CO;2-2

[50] Kambic, H.E. and C.A. McDevitt, *Spatial organization of types I and II collagen in the canine meniscus.* J Orthop Res, 2005. **23**(1): p. 142-9. DOI: 10.1016/j.orthres.2004.06.016

[51] Verdonk, P.C., R.G. Forsyth, J. Wang, K.F. Almqvist, R. Verdonk, E.M. Veys, and G. Verbruggen, *Characterisation of human knee meniscus cell phenotype.* Osteoarthritis Cartilage, 2005. **13**(7): p. 548-60. DOI: 10.1016/j.joca.2005.01.010

[52] Valiyaveettil, M., J.S. Mort, and C.A. McDevitt, *The concentration, gene expression, and spatial distribution of aggrecan in canine articular cartilage, meniscus, and anterior and posterior cruciate ligaments: a new molecular distinction between hyaline cartilage and fibrocartilage in the knee joint.* Connect Tissue Res, 2005. **46**(2): p. 83-91. DOI: 10.1080/03008200590954113

[53] Upton, M.L., J. Chen, and L.A. Setton, *Region-specific constitutive gene expression in the adult porcine meniscus.* J Orthop Res, 2006. **24**(7): p. 1562-70. DOI: 10.1002/jor.20146

[54] Cao, M., M. Stefanovic-Racic, H.I. Georgescu, L.A. Miller, and C.H. Evans, *Generation of nitric oxide by lapine meniscal cells and its effect on matrix metabolism: stimulation of collagen production by arginine.* J Orthop Res, 1998. **16**(1): p. 104-11. DOI: 10.1002/jor.1100160118

[55] Upton, M.L., F. Guilak, T.A. Laursen, and L.A. Setton, *Finite element modeling predictions of region-specific cell-matrix mechanics in the meniscus.* Biomech Model Mechanobiol, 2006. **5**(2-3): p. 140-9. DOI: 10.1007/s10237-006-0031-4

[56] Upton, M.L., A. Hennerbichler, B. Fermor, F. Guilak, J.B. Weinberg, and L.A. Setton, *Biaxial strain effects on cells from the inner and outer regions of the meniscus.* Connect Tissue Res, 2006. **47**(4): p. 207-14. DOI: 10.1080/03008200600846663

[57] Upton, M.L., J. Chen, F. Guilak, and L.A. Setton, *Differential effects of static and dynamic compression on meniscal cell gene expression.* J Orthop Res, 2003. **21**(6): p. 963-9. DOI: 10.1016/S0736-0266(03)00063-9

[58] Fink, C., B. Fermor, J.B. Weinberg, D.S. Pisetsky, M.A. Misukonis, and F. Guilak, *The effect of dynamic mechanical compression on nitric oxide production in the meniscus.* Osteoarthritis Cartilage, 2001. **9**(5): p. 481-7. DOI: 10.1053/joca.2001.0415

[59] Djurasovic, M., J.W. Aldridge, R. Grumbles, M.P. Rosenwasser, D. Howell, and A. Ratcliffe, *Knee joint immobilization decreases aggrecan gene expression in the meniscus.* Am J Sports Med, 1998. **26**(3): p. 460-6.

[60] Vandermeer, R.D. and F.K. Cunningham, *Arthroscopic treatment of the discoid lateral meniscus: results of long-term follow-up.* Arthroscopy, 1989. **5**(2): p. 101-9.

[61] Washington, E.R., 3rd, L. Root, and U.C. Liener, *Discoid lateral meniscus in children. Long-term follow-up after excision.* J Bone Joint Surg Am, 1995. **77**(9): p. 1357-61.

[62] Bjelle, A., *Cartilage matrix in hereditary pyrophosphate arthropathy.* J Rheumatol, 1981. **8**(6): p. 959-64.

[63] DiCarlo, E.F., *Pathology of the Meniscus, in Knee Meniscus: Basic and Clinical Foundations,* V.C. Mow, S.P. Arnoczky, and D.W. Jackson, Editors. 1992, Raven Press: New York. p. 123-128.

[64] Hough, A.J., Jr. and R.J. Webber, *Pathology of the meniscus.* Clin Orthop Relat Res, 1990. (252): p. 32-40.

[65] DeHaven, K.E., *Meniscectomy Versus Repair: Clinical Experience, in Knee Meniscus: Basic and Clinical Foundations,* V.C. Mow, S.P. Arnoczky, and D.W. Jackson, Editors. 1992, Raven Press: New York. p. 132.

[66] Berthiaume, M.J., J.P. Raynauld, J. Martel-Pelletier, F. Labonte, G. Beaudoin, D.A. Bloch, D. Choquette, B. Haraoui, R.D. Altman, M. Hochberg, J.M. Meyer, G.A. Cline, and J.P. Pelletier, *Meniscal tear and extrusion are strongly associated with progression of symptomatic knee osteoarthritis as assessed by quantitative magnetic resonance imaging.* Ann Rheum Dis, 2005. **64**(4): p. 556-63. DOI: 10.1136/ard.2004.023796

[67] Fahmy, N.R., E.A. Williams, and J. Noble, *Meniscal pathology and osteoarthritis of the knee.* J Bone Joint Surg Br, 1983. **65**(1): p. 24-8.

[68] Lange, A.K., M.A. Fiatarone Singh, R.M. Smith, N. Foroughi, M.K. Baker, R. Shnier, and B. Vanwanseele, *Degenerative meniscus tears and mobility impairment in women with knee osteoarthritis.* Osteoarthritis Cartilage, 2007. **15**(6): p. 701-8. DOI: 10.1016/j.joca.2006.11.004

[69] Bamac, B., S. Ozdemir, H.T. Sarisoy, T. Colak, A. Ozbek, and G. Akansel, *Evaluation of medial and lateral meniscus thicknesses in early osteoarthritis of the knee with magnetic resonance imaging.* Saudi Med J, 2006. **27**(6): p. 854-7.

[70] Adams, M.E., M.E. Billingham, and H. Muir, *The glycosaminoglycans in menisci in experimental and natural osteoarthritis.* Arthritis Rheum, 1983. **26**(1): p. 69-76. DOI: 10.1002/art.1780260111

[71] Sugita, T., T. Kawamata, M. Ohnuma, Y. Yoshizumi, and K. Sato, *Radial displacement of the medial meniscus in varus osteoarthritis of the knee.* Clin Orthop Relat Res, 2001. (387): p. 171-7. DOI: 10.1097/00003086-200106000-00023

[72] Smillie, I.S., *The current pattern of the pathology of meniscus tears.* Proc R Soc Med, 1968. **61**(1): p. 44-5.

[73] Schlossberg, S., H. Umans, G. Flusser, G.S. Difelice, and D.B. Lerer, *Bucket handle tears of the medial meniscus: meniscal intrusion rather than meniscal extrusion.* Skeletal Radiol, 2007. **36**(1): p. 29-34. DOI: 10.1007/s00256-006-0183-4

[74] Choi, N.H. and B.N. Victoroff, *Anterior horn tears of the lateral meniscus in soccer players.* Arthroscopy, 2006. **22**(5): p. 484-8.

[75] Dandy, D.J., *The arthroscopic anatomy of symptomatic meniscal lesions.* J Bone Joint Surg Br, 1990. **72**(4): p. 628-33.

[76] Noyes, F.R., E.S. Grood, D.L. Butler, and M. Malek, *Clinical laxity tests and functional stability of the knee: biomechanical concepts.* Clin Orthop Relat Res, 1980. (146): p. 84-9.

[77] Roberts, D., G. Andersson, and T. Friden, *Knee joint proprioception in ACL-deficient knees is related to cartilage injury, laxity and age: a retrospective study of 54 patients.* Acta Orthop Scand, 2004. **75**(1): p. 78-83. DOI: 10.1080/00016470410001708160

[78] Allen, C.R., E.K. Wong, G.A. Livesay, M. Sakane, F.H. Fu, and S.L. Woo, *Importance of the medial meniscus in the anterior cruciate ligament-deficient knee.* J Orthop Res, 2000. **18**(1): p. 109-15. DOI: 10.1002/jor.1100180116

[79] Belzer, J.P. and W.D. Cannon, Jr., *Meniscus Tears: Treatment in the Stable and Unstable Knee.* J Am Acad Orthop Surg, 1993. **1**(1): p. 41-47.

[80] Moglo, K.E. and A. Shirazi-Adl, *Biomechanics of passive knee joint in drawer: load transmission in intact and ACL-deficient joints.* Knee, 2003. **10**(3): p. 265-76. DOI: 10.1016/S0968-0160(02)00135-7

[81] Meister, K., P.A. Indelicato, S. Spanier, J. Franklin, and J. Batts, *Histology of the torn meniscus: a comparison of histologic differences in meniscal tissue between tears in anterior cruciate ligament-intact and anterior cruciate ligament-deficient knees.* Am J Sports Med, 2004. **32**(6): p. 1479-83. DOI: 10.1177/0363546503262182

[82] Fairbank, T.J., *Knee joint changes after meniscectomy.* J Bone Joint Surg Am, 1948. **30-B**(4): p. 664-70.

[83] McGinity, J.B., L.F. Geuss, and R.A. Marvin, *Partial or total meniscectomy: a comparative analysis.* J Bone Joint Surg Am, 1977. **59**(6): p. 763-6.

[84] Allen, P.R., R.A. Denham, and A.V. Swan, *Late degenerative changes after meniscectomy. Factors affecting the knee after operation.* J Bone Joint Surg Br, 1984. **66**(5): p. 666-71.

[85] Berjon, J.J., L. Munuera, and M. Calvo, *Degenerative lesions in the articular cartilage after meniscectomy: preliminary experimental study in dogs.* J Trauma, 1991. **31**(3): p. 342-50. DOI: 10.1097/00005373-199103000-00006

[86] Jackson, J.P., *Degenerative changes in the knee after meniscectomy.* Br Med J, 1968. **2**(5604): p. 525-7.

[87] Moon, M.S. and I.S. Chung, *Degenerative changes after meniscectomy and meniscal regeneration.* Int Orthop, 1988. **12**(1): p. 17-9. DOI: 10.1007/BF00265736

[88] Messner, K. and J. Gao, *The menisci of the knee joint. Anatomical and functional characteristics, and a rationale for clinical treatment.* J Anat, 1998. **193**(Pt 2): p. 161-78. DOI: 10.1046/j.1469-7580.1998.19320161.x

[89] Goodfellow, J., *He who hesitates is saved.* J Bone Joint Surg Br, 1980. **62-B**(1): p. 1-2.

[90] Adams, S.B., Jr., M.A. Randolph, and T.J. Gill, *Tissue engineering for meniscus repair.* J Knee Surg, 2005. **18**(1): p. 25-30.

[91] Hoben, G.M. and K.A. Athanasiou, *Meniscal repair with fibrocartilage engineering.* Sports Med Arthrosc, 2006. **14**(3): p. 129-37. DOI: 10.1097/00132585-200609000-00004

[92] Arnoczky, S.P., *Gross and Vascular Anatomy of the Meniscus and Its Role in Meniscal Healing, Regeneration, and Remodeling, in Knee Meniscus: Basic and Clinical Foundations, V.C. Mow, S.P. Arnoczky, and D.W. Jackson, Editors. 1992,* Raven Press: New York. p. 6-12.

[93] Arnoczky, S.P. and R.F. Warren, *The microvasculature of the meniscus and its response to injury. An experimental study in the dog.* Am J Sports Med, 1983. **11**(3): p. 131-41. DOI: 10.1177/036354658301100305

[94] Mow, V.C., S.P. Arnoczky, and D.W. Jackson, *Knee Meniscus: Basic and Clinical Foundations.* 1992. p. 6-12.

[95] McAndrews, P.T. and S.P. Arnoczky, *Meniscal repair enhancement techniques.* Clin Sports Med, 1996. **15**(3): p. 499-510.

[96] Zhang, Z. and J.A. Arnold, *Trephination and suturing of avascular meniscal tears: a clinical study of the trephination procedure.* Arthroscopy, 1996. **12**(6): p. 726-31.

[97] Fox, J.M., K.G. Rintz, and R.D. Ferkel, *Trephination of incomplete meniscal tears.* Arthroscopy, 1993. **9**(4): p. 451-5.

[98] Zhang, Z.N., K.Y. Tu, Y.K. Xu, W.M. Zhang, Z.T. Liu, and S.H. Ou, *Treatment of longitudinal injuries in avascular area of meniscus in dogs by trephination.* Arthroscopy, 1988. **4**(3): p. 151-9.

[99] Zhang, Z., J.A. Arnold, T. Williams, and B. McCann, *Repairs by trephination and suturing of longitudinal injuries in the avascular area of the meniscus in goats.* Am J Sports Med, 1995. **23**(1): p. 35-41. DOI: 10.1177/036354659502300106

[100] Noble, J. and D.L. Hamblen, *The pathology of the degenerate meniscus lesion.* J Bone Joint Surg Br, 1975. **57**(2): p. 180-6.

[101] Uysal, M., S. Akpinar, F. Bolat, N. Cekin, M. Cinar, and N. Cesur, *Apoptosis in the traumatic and degenerative tears of human meniscus.* Knee Surg Sports Traumatol Arthrosc, 2008. **16**(7): p. 666-9. DOI: 10.1007/s00167-008-0536-8

[102] Roeddecker, K., U. Muennich, and M. Nagelschmidt, *Meniscal healing: a biomechanical study.* J Surg Res, 1994. **56**(1): p. 20-7. DOI: 10.1006/jsre.1994.1004

[103] Aufderheide, A.C. and K.A. Athanasiou, *Assessment of a bovine co-culture, scaffold-free method for growing meniscus-shaped constructs.* Tissue Eng, 2007. **13**(9): p. 2195-205. DOI: 10.1089/ten.2006.0291

[104] Hu, J.C. and K.A. Athanasiou, *A self-assembling process in articular cartilage tissue engineering.* Tissue Eng, 2006. **12**(4): p. 969-79. DOI: 10.1089/ten.2006.12.969

[105] Lal, B., J. Viola, D. Hicks, and O. Grad, *Emergence and Evolution of a Shared Concept, in The Emergence of Tissue Engineering as a Research Field.* 2003, Prepared for the National Science Foundation: Arlington, VA.

[106] Vacanti, C.A., *The history of tissue engineering.* J Cell Mol Med, 2006. **10**(3): p. 569-76. DOI: 10.1111/j.1582-4934.2006.tb00421.x

[107] Butler, D.L., S.A. Goldstein, and F. Guilak, *Functional tissue engineering: the role of biomechanics.* J Biomech Eng, 2000. **122**(6): p. 570-5. DOI: 10.1115/1.1318906

[108] Goldstein, S.A., *Tissue engineering: functional assessment and clinical outcome.* Ann N Y Acad Sci, 2002. **961**: p. 183-92.

[109] Guilak, F., *Functional tissue engineering: the role of biomechanics in reparative medicine.* Ann N Y Acad Sci, 2002. **961**: p. 193-5.

[110] Guilak, F., D.L. Butler, and S.A. Goldstein, *Functional tissue engineering: the role of biomechanics in articular cartilage repair.* Clin Orthop Relat Res, 2001. (391 Suppl): p. S295-305. DOI: 10.1097/00003086-200110001-00027

[111] Mauck, R.L., M.A. Soltz, C.C. Wang, D.D. Wong, P.H. Chao, W.B. Valhmu, C.T. Hung, and G.A. Ateshian, *Functional tissue engineering of articular cartilage through dynamic loading of chondrocyte-seeded agarose gels.* J Biomech Eng, 2000. **122**(3): p. 252-60. DOI: 10.1115/1.429656

[112] Hoben, G.M. and K.A. Athanasiou, *Creating a spectrum of fibrocartilages through different cell sources and biochemical stimuli.* Biotechnol Bioeng, 2008. **100**(3): p. 587-98. DOI: 10.1002/bit.21768

[113] Hoben, G.M., J.C. Hu, R.A. James, and K.A. Athanasiou, *Self-assembly of fibrochondrocytes and chondrocytes for tissue engineering of the knee meniscus.* Tissue Eng, 2007. **13**(5): p. 939-46. DOI: 10.1089/ten.2006.0116

[114] Setton, L.A., F. Guilak, E.W. Hsu, and T.P. Vail, *Biomechanical factors in tissue engineered meniscal repair.* Clin Orthop Relat Res, 1999. (367 Suppl): p. S254-72. DOI: 10.1097/00003086-199910001-00025

[115] Mikic, B., T.L. Johnson, A.B. Chhabra, B.J. Schalet, M. Wong, and E.B. Hunziker, *Differential effects of embryonic immobilization on the development of fibrocartilaginous skeletal elements.* J Rehabil Res Dev, 2000. **37**(2): p. 127-33.

[116] Ochi, M., T. Kanda, Y. Sumen, and Y. Ikuta, *Changes in the permeability and histologic findings of rabbit menisci after immobilization.* Clin Orthop Relat Res, 1997. (334): p. 305-15. DOI: 10.1097/00003086-199701000-00040

[117] Natsu-Ume, T., T. Majima, C. Reno, N.G. Shrive, C.B. Frank, and D.A. Hart, *Menisci of the rabbit knee require mechanical loading to maintain homeostasis: cyclic hydrostatic compression in vitro prevents derepression of catabolic genes.* J Orthop Sci, 2005. **10**(4): p. 396-405. DOI: 10.1007/s00776-005-0912-x

[118] Buschmann, M.D., Y.A. Gluzband, A.J. Grodzinsky, and E.B. Hunziker, *Mechanical compression modulates matrix biosynthesis in chondrocyte/agarose culture.* J Cell Sci, 1995. **108**(Pt 4): p. 1497-508.

[119] Hung, C.T., R.L. Mauck, C.C. Wang, E.G. Lima, and G.A. Ateshian, *A paradigm for functional tissue engineering of articular cartilage via applied physiologic deformational loading.* Ann Biomed Eng, 2004. **32**(1): p. 35-49. DOI: 10.1023/B:ABME.0000007789.99565.42

[120] Mauck, R.L., S.L. Seyhan, G.A. Ateshian, and C.T. Hung, *Influence of seeding density and dynamic deformational loading on the developing structure/function relationships of chondrocyte-seeded agarose hydrogels.* Ann Biomed Eng, 2002. **30**(8): p. 1046-56. DOI: 10.1114/1.1512676

[121] Sah, R.L., Y.J. Kim, J.Y. Doong, A.J. Grodzinsky, A.H. Plaas, and J.D. Sandy, *Biosynthetic response of cartilage explants to dynamic compression.* J Orthop Res, 1989. **7**(5): p. 619-36. DOI: 10.1002/jor.1100070502

[122] Meyer, U., A. Buchter, N. Nazer, and H.P. Wiesmann, *Design and performance of a bioreactor system for mechanically promoted three-dimensional tissue engineering.* Br J Oral Maxillofac Surg, 2006. **44**(2): p. 134-40. DOI: 10.1016/j.bjoms.2005.05.001

[123] Aufderheide, A.C. and K.A. Athanasiou, *A direct compression stimulator for articular cartilage and meniscal explants.* Ann Biomed Eng, 2006. **34**(9): p. 1463-74. DOI: 10.1007/s10439-006-9157-x

[124] McHenry, J.A., B. Zielinska, and T.L. Donahue, *Proteoglycan breakdown of meniscal explants following dynamic compression using a novel bioreactor.* Ann Biomed Eng, 2006. **34**(11): p. 1758-66. DOI: 10.1007/s10439-006-9178-5

[125] Smith, R.L., J. Tin, M.C. Trindade, J. Shida, G. Kajiyama, T. Vu, A.R. Hoffman, M.C. van der Meulen, S.B. Goodman, D.J. Schurman, and D.R. Carter, *Time-dependent effects of intermittent hydrostatic pressure on articular chondrocyte type II collagen and aggrecan mRNA expression.* J Rehabil Res Dev, 2000. **37**(2): p. 153-61.

[126] Carver, S.E. and C.A. Heath, *Semi-continuous perfusion system for delivering intermittent physiological pressure to regenerating cartilage.* Tissue Eng, 1999. **5**(1): p. 1-11. DOI: 10.1089/ten.1999.5.1

[127] Carver, S.E. and C.A. Heath, *Increasing extracellular matrix production in regenerating cartilage with intermittent physiological pressure.* Biotechnol Bioeng, 1999. **62**(2): p. 166-74. DOI: 10.1002/(SICI)1097-0290(19990120)62:2<166::AID-BIT6>3.3.CO;2-B

[128] Elder, B.D. and K.A. Athanasiou, *Effects of Temporal Hydrostatic Pressure on Tissue-Engineered Bovine Articular Cartilage Constructs.* Tissue Eng Part A, 2008. DOI: 10.1089/ten.tea.2008.0200

[129] Elder, B.D. and K.A. Athanasiou, *Synergistic and additive effects of hydrostatic pressure and growth factors on tissue formation.* PLoS ONE, 2008. **3**(6): p. e2341. DOI: 10.1371/journal.pone.0002341

[130] Gunja, N.J., R.K. Uthamanthil, and K.A. Athanasiou, *Effects of TGF-beta1 and hydrostatic pressure on meniscus cell-seeded scaffolds.* Biomaterials, 2009. **30**(4): p. 565-73.

[131] Hu, J.C. and K.A. Athanasiou, *The effects of intermittent hydrostatic pressure on self-assembled articular cartilage constructs.* Tissue Eng, 2006. **12**(5): p. 1337-44. DOI: 10.1089/ten.2006.12.1337

[132] Almarza, A.J. and K.A. Athanasiou, *Effects of hydrostatic pressure on TMJ disc cells.* Tissue Eng, 2006. **12**(5): p. 1285-94. DOI: 10.1089/ten.2006.12.1285

[133] Mahmoudifar, N. and P.M. Doran, *Tissue engineering of human cartilage in bioreactors using single and composite cell-seeded scaffolds.* Biotechnol Bioeng, 2005. **91**(3): p. 338-55. DOI: 10.1002/bit.20490

[134] Nagel-Heyer, S., C. Goepfert, F. Feyerabend, J.P. Petersen, P. Adamietz, N.M. Meenen, and R. Portner, *Bioreactor cultivation of three-dimensional cartilage-carrier-constructs.* Bioprocess Biosyst Eng, 2005. **27**(4): p. 273-80. DOI: 10.1007/s00449-005-0419-z

[135] Neves, A.A., N. Medcalf, and K.M. Brindle, *Tissue engineering of meniscal cartilage using perfusion culture.* Ann N Y Acad Sci, 2002. **961**: p. 352-5.

[136] Seidel, J.O., M. Pei, M.L. Gray, R. Langer, L.E. Freed, and G. Vunjak-Novakovic, *Long-term culture of tissue engineered cartilage in a perfused chamber with mechanical stimulation.* Biorheology, 2004. **41**(3-4): p. 445-58.

[137] Darling, E.M. and K.A. Athanasiou, *Articular cartilage bioreactors and bioprocesses.* Tissue Eng, 2003. **9**(1): p. 9-26. DOI: 10.1089/107632703762687492

[138] Vunjak-Novakovic, G., I. Martin, B. Obradovic, S. Treppo, A.J. Grodzinsky, R. Langer, and L.E. Freed, *Bioreactor cultivation conditions modulate the composition and mechanical properties of tissue-engineered cartilage.* J Orthop Res, 1999. **17**(1): p. 130-8. DOI: 10.1002/jor.1100170119

[139] Gemmiti, C.V. and R.E. Guldberg, *Fluid flow increases type II collagen deposition and tensile mechanical properties in bioreactor-grown tissue-engineered cartilage.* Tissue Eng, 2006. **12**(3): p. 469-79. DOI: 10.1089/ten.2006.12.469

[140] Freed, L.E., A.P. Hollander, I. Martin, J.R. Barry, R. Langer, and G. Vunjak-Novakovic, *Chondrogenesis in a cell-polymer-bioreactor system.* Exp Cell Res, 1998. **240**(1): p. 58-65. DOI: 10.1006/excr.1998.4010

[141] Aufderheide, A.C. and K.A. Athanasiou, *Comparison of scaffolds and culture conditions for tissue engineering of the knee meniscus.* Tissue Eng, 2005. **11**(7-8): p. 1095-104. DOI: 10.1089/ten.2005.11.1095

[142] Bueno, E.M., B. Bilgen, and G.A. Barabino, *Wavy-walled bioreactor supports increased cell proliferation and matrix deposition in engineered cartilage constructs.* Tissue Eng, 2005. **11**(11-12): p. 1699-709. DOI: 10.1089/ten.2005.11.1699

[143] Freed, L.E., N. Pellis, N. Searby, J. de Luis, C. Preda, J. Bordonaro, and G. Vunjak-Novakovic, *Microgravity cultivation of cells and tissues.* Gravit Space Biol Bull, 1999. **12**(2): p. 57-66.

[144] Bilgen, B., K. Uygun, E.M. Bueno, P. Sucosky, and G.A. Barabino, *Tissue Growth Modeling in a Wavy-Walled Bioreactor.* Tissue Eng Part A, 2008. DOI: 10.1089/ten.tea.2008.0078

[145] Bilgen, B., P. Sucosky, G.P. Neitzel, and G.A. Barabino, *Flow characterization of a wavy-walled bioreactor for cartilage tissue engineering.* Biotechnol Bioeng, 2006. **95**(6): p. 1009-22. DOI: 10.1002/bit.20775

[146] Bilgen, B. and G.A. Barabino, *Location of scaffolds in bioreactors modulates the hydrodynamic environment experienced by engineered tissues.* Biotechnol Bioeng, 2007. **98**(1): p. 282-94. DOI: 10.1002/bit.21385

[147] Yang, K.H., J. Parvizi, S.J. Wang, D.G. Lewallen, R.R. Kinnick, J.F. Greenleaf, and M.E. Bolander, *Exposure to low-intensity ultrasound increases aggrecan gene expression in a rat femur fracture model.* J Orthop Res, 1996. **14**(5): p. 802-9. DOI: 10.1002/jor.1100140518

[148] Parvizi, J., C.C. Wu, D.G. Lewallen, J.F. Greenleaf, and M.E. Bolander, *Low-intensity ultrasound stimulates proteoglycan synthesis in rat chondrocytes by increasing aggrecan gene expression.* J Orthop Res, 1999. **17**(4): p. 488-94. DOI: 10.1002/jor.1100170405

[149] Parvizi, J., V. Parpura, J.F. Greenleaf, and M.E. Bolander, *Calcium signaling is required for ultrasound-stimulated aggrecan synthesis by rat chondrocytes.* J Orthop Res, 2002. **20**(1): p. 51-7. DOI: 10.1016/S0736-0266(01)00069-9

[150] Hsu, S.H., C.C. Kuo, S.W. Whu, C.H. Lin, and C.L. Tsai, *The effect of ultrasound stimulation versus bioreactors on neocartilage formation in tissue engineering scaffolds seeded with human chondrocytes in vitro.* Biomol Eng, 2006. **23**(5): p. 259-64. DOI: 10.1016/j.bioeng.2006.05.029

[151] Duda, G.N., A. Kliche, R. Kleemann, J.E. Hoffmann, M. Sittinger, and A. Haisch, *Does low-intensity pulsed ultrasound stimulate maturation of tissue-engineered cartilage?* J Biomed Mater Res B Appl Biomater, 2004. **68**(1): p. 21-8. DOI: 10.1002/jbm.b.10075

[152] Stoddart, M.J., L. Ettinger, and H.J. Hauselmann, *Enhanced matrix synthesis in de novo, scaffold free cartilage-like tissue subjected to compression and shear.* Biotechnol Bioeng, 2006. **95**(6): p. 1043-51. DOI: 10.1002/bit.21052

[153] Wimmer, M.A., S. Grad, T. Kaup, M. Hanni, E. Schneider, S. Gogolewski, and M. Alini, *Tribology approach to the engineering and study of articular cartilage.* Tissue Eng, 2004. **10**(9-10): p. 1436-45. DOI: 10.1089/ten.2004.10.1436

[154] Bhargava, M.M., E.T. Attia, G.A. Murrell, M.M. Dolan, R.F. Warren, and J.A. Hannafin, *The effect of cytokines on the proliferation and migration of bovine meniscal cells.* Am J Sports Med, 1999. **27**(5): p. 636-43.

[155] Pangborn, C.A. and K.A. Athanasiou, *Growth factors and fibrochondrocytes in scaffolds.* J Orthop Res, 2005. **23**(5): p. 1184-90. DOI: 10.1016/j.orthres.2005.01.019

[156] Kawamura, K., C.R. Chu, S. Sobajima, P.D. Robbins, F.H. Fu, N.J. Izzo, and C. Niyibizi, *Adenoviral-mediated transfer of TGF-beta1 but not IGF-1 induces chondrogenic differentiation of human mesenchymal stem cells in pellet cultures.* Exp Hematol, 2005. **33**(8): p. 865-72.

[157] Sekiya, I., J.T. Vuoristo, B.L. Larson, and D.J. Prockop, *In vitro cartilage formation by human adult stem cells from bone marrow stroma defines the sequence of cellular and molecular events during chondrogenesis.* Proc Natl Acad Sci U S A, 2002. **99**(7): p. 4397-402. DOI: 10.1073/pnas.052716199

[158] Worster, A.A., B.D. Brower-Toland, L.A. Fortier, S.J. Bent, J. Williams, and A.J. Nixon, *Chondrocytic differentiation of mesenchymal stem cells sequentially exposed to transforming growth factor-beta1 in monolayer and insulin-like growth factor-I in a three-dimensional matrix.* J Orthop Res, 2001. **19**(4): p. 738-49. DOI: 10.1016/S0736-0266(00)00054-1

[159] Hoben, G.M., V.P. Willard, and K.A. Athanasiou, *Fibrochondrogenesis of hESCs: Growth factor combinations and co-cultures.* Stem Cells Dev, 2008.

[160] Hoben, G.M., E.J. Koay, and K.A. Athanasiou, *Fibrochondrogenesis in two embryonic stem cell lines: effects of differentiation timelines.* Stem Cells, 2008. **26**(2): p. 422-30. DOI: 10.1634/stemcells.2007-0641

[161] Van Osch, G.J., E.W. Mandl, H. Jahr, W. Koevoet, G. Nolst-Trenite, and J.A. Verhaar, *Considerations on the use of ear chondrocytes as donor chondrocytes for cartilage tissue engineering.* Biorheology, 2004. **41**(3-4): p. 411-21.

[162] Deng, Y., J.C. Hu, and K.A. Athanasiou, *Isolation and chondroinduction of a dermis-isolated, aggrecan-sensitive subpopulation with high chondrogenic potential.* Arthritis Rheum, 2007. **56**(1): p. 168-76. DOI: 10.1002/art.22300

[163] Ikeda, T., S. Kamekura, A. Mabuchi, I. Kou, S. Seki, T. Takato, K. Nakamura, H. Kawaguchi, S. Ikegawa, and U.I. Chung, *The combination of SOX5, SOX6, and SOX9 (the SOX trio) provides signals sufficient for induction of permanent cartilage.* Arthritis Rheum, 2004. **50**(11): p. 3561-73. DOI: 10.1002/art.20611

[164] Mizuno, S. and J. Glowacki, *Chondroinduction of human dermal fibroblasts by demineralized bone in three-dimensional culture.* Exp Cell Res, 1996. **227**(1): p. 89-97. DOI: 10.1006/excr.1996.0253

[165] Betre, H., S.R. Ong, F. Guilak, A. Chilkoti, B. Fermor, and L.A. Setton, *Chondrocytic differentiation of human adipose-derived adult stem cells in elastin-like polypeptide.* Biomaterials, 2006. **27**(1): p. 91-9. DOI: 10.1016/j.biomaterials.2005.05.071

[166] Estes, B.T., A.W. Wu, and F. Guilak, *Potent induction of chondrocytic differentiation of human adipose-derived adult stem cells by bone morphogenetic protein 6.* Arthritis Rheum, 2006. **54**(4): p. 1222-32. DOI: 10.1002/art.21779

[167] Gimble, J. and F. Guilak, *Adipose-derived adult stem cells: isolation, characterization, and differentiation potential.* Cytotherapy, 2003. **5**(5): p. 362-9. DOI: 10.1080/14653240310003026

[168] Park, Y., M. Sugimoto, A. Watrin, M. Chiquet, and E.B. Hunziker, *BMP-2 induces the expression of chondrocyte-specific genes in bovine synovium-derived progenitor cells cultured in three-dimensional alginate hydrogel.* Osteoarthritis Cartilage, 2005. **13**(6): p. 527-36. DOI: 10.1016/j.joca.2005.02.006

[169] Collier, S. and P. Ghosh, *Effects of transforming growth factor beta on proteoglycan synthesis by cell and explant cultures derived from the knee joint meniscus.* Osteoarthritis Cartilage, 1995. **3**(2): p. 127-38. DOI: 10.1016/S1063-4584(05)80045-7

[170] Spindler, K.P., C.E. Mayes, R.R. Miller, A.K. Imro, and J.M. Davidson, *Regional mitogenic response of the meniscus to platelet-derived growth factor (PDGF-AB).* J Orthop Res, 1995. **13**(2): p. 201-7. DOI: 10.1002/jor.1100130208

[171] Steinert, A.F., G.D. Palmer, R. Capito, J.G. Hofstaetter, C. Pilapil, S.C. Ghivizzani, M. Spector, and C.H. Evans, *Genetically enhanced engineering of meniscus tissue using ex vivo delivery of transforming growth factor-beta 1 complementary deoxyribonucleic acid.* Tissue Eng, 2007. **13**(9): p. 2227-37. DOI: 10.1089/ten.2006.0270

[172] Marsano, A., G. Vunjak-Novakovic, and I. Martin, *Towards tissue engineering of meniscus substitutes: selection of cell source and culture environment.* Conf Proc IEEE Eng Med Biol Soc, 2006. **1**: p. 3656-8. DOI: 10.1109/IEMBS.2006.259748

[173] Marsano, A., D. Wendt, R. Raiteri, R. Gottardi, M. Stolz, D. Wirz, A.U. Daniels, D. Salter, M. Jakob, T.M. Quinn, and I. Martin, *Use of hydrodynamic forces to engineer cartilaginous tissues resembling the non-uniform structure and function of meniscus.* Biomaterials, 2006. **27**(35): p. 5927-34. DOI: 10.1016/j.biomaterials.2006.08.020

[174] Maier, D., K. Braeun, E. Steinhauser, P. Ueblacker, M. Oberst, P.C. Kreuz, N. Roos, V. Martinek, and A.B. Imhoff, *In vitro analysis of an allogenic scaffold for tissue-engineered meniscus replacement.* J Orthop Res, 2007. **25**(12): p. 1598-608. DOI: 10.1002/jor.20405

[175] Angele, P., B. Johnstone, R. Kujat, J. Zellner, M. Nerlich, V. Goldberg, and J. Yoo, *Stem cell based tissue engineering for meniscus repair.* J Biomed Mater Res A, 2008. **85**(2): p. 445-55.

[176] Kang, S.W., S.M. Son, J.S. Lee, E.S. Lee, K.Y. Lee, S.G. Park, J.H. Park, and B.S. Kim, *Regeneration of whole meniscus using meniscal cells and polymer scaffolds in a rabbit total meniscectomy model.* J Biomed Mater Res A, 2006. **78**(3): p. 659-71.

[177] Testa Pezzin, A.P., T.P. Cardoso, M. do Carmo Alberto Rincon, C.A. de Carvalho Zavaglia, and E.A. de Rezende Duek, *Bioreabsorbable polymer scaffold as temporary meniscal prosthesis.* Artif Organs, 2003. **27**(5): p. 428-31. DOI: 10.1046/j.1525-1594.2003.07251.x

[178] Sonoda, M., F.L. Harwood, M.E. Amiel, H. Moriya, M. Temple, D.G. Chang, L.M. Lottman, R.L. Sah, and D. Amiel, *The effects of hyaluronan on tissue healing after meniscus injury and repair in a rabbit model.* Am J Sports Med, 2000. **28**(1): p. 90-7.

[179] Walsh, C.J., D. Goodman, A.I. Caplan, and V.M. Goldberg, *Meniscus regeneration in a rabbit partial meniscectomy model.* Tissue Eng, 1999. **5**(4): p. 327-37. DOI: 10.1089/ten.1999.5.327

[180] Suzuki, Y., N. Takeuchi, Y. Sagehashi, T. Yamaguchi, H. Itoh, and H. Iwata, *Effects of hyaluronic acid on meniscal injury in rabbits.* Arch Orthop Trauma Surg, 1998. **117**(6-7): p. 303-6. DOI: 10.1007/s004020050255

[181] Isoda, K. and S. Saito, *In vitro and in vivo fibrochondrocyte growth behavior in fibrin gel: an immunohistochemical study in the rabbit.* Am J Knee Surg, 1998. **11**(4): p. 209-16.

[182] Bradley, M.P., P.D. Fadale, M.J. Hulstyn, W.R. Muirhead, and J.T. Lifrak, *Porcine small intestine submucosa for repair of goat meniscal defects.* Orthopedics, 2007. **30**(8): p. 650-6.

[183] Port, J., D.W. Jackson, T.Q. Lee, and T.M. Simon, *Meniscal repair supplemented with exogenous fibrin clot and autogenous cultured marrow cells in the goat model.* Am J Sports Med, 1996. **24**(4): p. 547-55. DOI: 10.1177/036354659602400422

[184] Martinek, V., P. Ueblacker, K. Braun, S. Nitschke, R. Mannhardt, K. Specht, B. Gansbacher, and A.B. Imhoff, *Second generation of meniscus transplantation: in-vivo study with tissue engineered meniscus replacement.* Arch Orthop Trauma Surg, 2006. **126**(4): p. 228-34. DOI: 10.1007/s00402-005-0025-1

[185] Burger, C., K. Kabir, C. Rangger, M. Mueller, T. Minor, and R.H. Tolba, *Polylactide (LTS) causes less inflammation response than polydioxanone (PDS): a meniscus repair model in sheep.* Arch Orthop Trauma Surg, 2006. **126**(10): p. 695-705. DOI: 10.1007/s00402-006-0207-5

[186] Nabeshima, Y., M. Kurosaka, S. Yoshiya, and K. Mizuno, *Effect of fibrin glue and endothelial cell growth factor on the early healing response of the transplanted allogenic meniscus: a pilot study.* Knee Surg Sports Traumatol Arthrosc, 1995. **3**(1): p. 34-8. DOI: 10.1007/BF01553523

[187] Hashimoto, J., M. Kurosaka, S. Yoshiya, and K. Hirohata, *Meniscal repair using fibrin sealant and endothelial cell growth factor. An experimental study in dogs.* Am J Sports Med, 1992. **20**(5): p. 537-41. DOI: 10.1177/036354659202000509

[188] Peretti, G.M., T.J. Gill, J.W. Xu, M.A. Randolph, K.R. Morse, and D.J. Zaleske, *Cell-based therapy for meniscal repair: a large animal study.* Am J Sports Med, 2004. **32**(1): p. 146-58. DOI: 10.1177/0095399703258790

[189] Steadman, J.R. and W.G. Rodkey, *Tissue-engineered collagen meniscus implants: 5- to 6-year feasibility study results.* Arthroscopy, 2005. **21**(5): p. 515-25.

[190] Reguzzoni, M., A. Manelli, M. Ronga, M. Raspanti, and F.A. Grassi, *Histology and ultrastructure of a tissue-engineered collagen meniscus before and after implantation.* J Biomed Mater Res B Appl Biomater, 2005. **74**(2): p. 808-16.

[191] Ibarra, C., C. Jannetta, C.A. Vacanti, Y. Cao, T.H. Kim, J. Upton, and J.P. Vacanti, *Tissue engineered meniscus: a potential new alternative to allogeneic meniscus transplantation.* Transplant Proc, 1997. **29**(1-2): p. 986-8. DOI: 10.1016/S0041-1345(96)00337-5

[192] Hidaka, C., C. Ibarra, J.A. Hannafin, P.A. Torzilli, M. Quitoriano, S.S. Jen, R.F. Warren, and R.G. Crystal, *Formation of vascularized meniscal tissue by combining gene therapy with tissue engineering.* Tissue Eng, 2002. **8**(1): p. 93-105. DOI: 10.1089/107632702753503090

[193] Rodkey, W.G., K.E. DeHaven, W.H. Montgomery, 3rd, C.L. Baker, Jr., C.L. Beck, Jr., S.E. Hormel, J.R. Steadman, B.J. Cole, and K.K. Briggs, *Comparison of the collagen meniscus implant with partial meniscectomy. A prospective randomized trial.* J Bone Joint Surg Am, 2008. **90**(7): p. 1413-26. DOI: 10.2106/JBJS.G.00656

[194] Boyd, K.T. and P.T. Myers, *Meniscus preservation; rationale, repair techniques and results.* Knee, 2003. **10**(1): p. 1-11. DOI: 10.1016/S0968-0160(02)00147-3

[195] Espejo-Baena, A., A. Figueroa-Mata, J. Serrano-Fernandez, and F. de la Torre-Solis, *All-inside suture technique using anterior portals in posterior horn tears of lateral meniscus.* Arthroscopy, 2008. **24**(3): p. 369 e1-4.

[196] Hantes, M.E., V.C. Zachos, S.E. Varitimidis, Z.H. Dailiana, T. Karachalios, and K.N. Malizos, *Arthroscopic meniscal repair: a comparative study between three different surgical techniques.* Knee Surg Sports Traumatol Arthrosc, 2006. **14**(12): p. 1232-7. DOI: 10.1007/s00167-006-0094-x

[197] Kim, S.J., K.A. Jung, J.M. Kim, J.D. Kwun, S.H. Baek, and J.N. Han, *Arthroscopic all-inside repair of tears of the anterior horn of the lateral meniscus.* Arthroscopy, 2005. **21**(11): p. 1399.

[198] Fukushima, K., T. Okano, S. Negishi, T. Horaguchi, K. Sato, A. Saito, and J. Ryu, *New meniscus repair by an all-inside knot suture technique.* Arthroscopy, 2005. **21**(6): p. 768.

[199] Cho, J.H., *Arthroscopic all-inside repair of anterior horn tears of the lateral meniscus using a spinal needle.* Knee Surg Sports Traumatol Arthrosc, 2008. **16**(7): p. 683-6. DOI: 10.1007/s00167-007-0483-9

[200] Gill, S.S. and D.R. Diduch, *Outcomes after meniscal repair using the meniscus arrow in knees undergoing concurrent anterior cruciate ligament reconstruction.* Arthroscopy, 2002. **18**(6): p. 569-77.

[201] Albrecht-Olsen, P., T. Lind, G. Kristensen, and B. Falkenberg, *Failure strength of a new meniscus arrow repair technique: biomechanical comparison with horizontal suture.* Arthroscopy, 1997. **13**(2): p. 183-7.

[202] Kurzweil, P.R., C.D. Tifford, and E.M. Ignacio, *Unsatisfactory clinical results of meniscal repair using the meniscus arrow.* Arthroscopy, 2005. **21**(8): p. 905.

[203] Koukoulias, N., S. Papastergiou, K. Kazakos, G. Poulios, and K. Parisis, *Clinical results of meniscus repair with the meniscus arrow: a 4- to 8-year follow-up study.* Knee Surg Sports Traumatol Arthrosc, 2007. **15**(2): p. 133-7. DOI: 10.1007/s00167-006-0141-7

[204] Lee, G.P. and D.R. Diduch, *Deteriorating outcomes after meniscal repair using the Meniscus Arrow in knees undergoing concurrent anterior cruciate ligament reconstruction: increased failure rate with long-term follow-up.* Am J Sports Med, 2005. **33**(8): p. 1138-41. DOI: 10.1177/0363546505275348

[205] Jones, H.P., M.J. Lemos, R.M. Wilk, P.M. Smiley, R. Gutierrez, and A.A. Schepsis, *Two-year follow-up of meniscal repair using a bioabsorbable arrow.* Arthroscopy, 2002. **18**(1): p. 64-9.

[206] Albrecht-Olsen, P., G. Kristensen, P. Burgaard, U. Joergensen, and C. Toerholm, *The arrow versus horizontal suture in arthroscopic meniscus repair. A prospective randomized study with arthroscopic evaluation.* Knee Surg Sports Traumatol Arthrosc, 1999. **7**(5): p. 268-73. DOI: 10.1007/s001670050162

[207] Hantes, M.E., E.S. Kotsovolos, D.S. Mastrokalos, J. Ammenwerth, and H.H. Paessler, *Arthroscopic meniscal repair with an absorbable screw: results and surgical technique.* Knee Surg Sports Traumatol Arthrosc, 2005. **13**(4): p. 273-9. DOI: 10.1007/s00167-004-0527-3

[208] Siebold, R., C. Dehler, L. Boes, and A. Ellermann, *Arthroscopic all-inside repair using the Meniscus Arrow: long-term clinical follow-up of 113 patients.* Arthroscopy, 2007. **23**(4): p. 394-9.

[209] Garrett, J.C. and R.N. Steensen, *Meniscal transplantation in the human knee: a preliminary report.* Arthroscopy, 1991. **7**(1): p. 57-62.

[210] Stone, K.R., A.W. Walgenbach, T.J. Turek, A. Freyer, and M.D. Hill, *Meniscus allograft survival in patients with moderate to severe unicompartmental arthritis: a 2- to 7-year follow-up.* Arthroscopy, 2006. **22**(5): p. 469-78.

[211] McDermott, I. and N.P. Thomas, *Human meniscal allograft transplantation.* Knee, 2006. **13**(1): p. 69-71. DOI: 10.1016/j.knee.2005.08.006

[212] Shelbourne, K.D. and J. Heinrich, *The long-term evaluation of lateral meniscus tears left in situ at the time of anterior cruciate ligament reconstruction.* Arthroscopy, 2004. **20**(4): p. 346-51.

[213] Elder, B.D. and K.A. Athanasiou, *Systematic assessment of growth factor treatment on biochemical and biomechanical properties of engineered articular cartilage constructs.* Osteoarthritis Cartilage, 2009. **17**(1): p. 114-23. DOI: 10.1016/j.joca.2008.05.006

Acknowledgments

We would like to thank Ms. Teresa Sanchez Adams for proofreading the text and providing useful comments and suggestions. We would also like to extend our gratitude to the many graduate students that have worked on meniscus tissue engineering in Professor Athanasiou's lab laying the groundwork for this book, including Dr. Mark Sweigart, Dr. Adam Aufderheide, Ms. Christie Pangborn, Dr. Gwendolyn Hoben, Dr. Jerry Hu, Dr. Najmuddin Gunja, and Mr. Dan Huey. Finally, we would like to thank the National Institutes of Health, specifically the National Institute of Arthritis and Musculoskeletal and Skin Diseases, for their support of our group's knee meniscus tissue engineering efforts.

About the Authors

K. A. ATHANASIOU

K. A. Athanasiou is the Karl F. Hasselmann professor of bioengineering at Rice University and an adjunct professor of orthopedics and oral and maxillofacial surgery at the University of Texas. He also heads the Musculoskeletal Bioengineering Laboratory at Rice University. He holds a Ph.D. in mechanical engineering (bioengineering) from Columbia University.

JOHANNAH SANCHEZ-ADAMS

Johannah Sanchez-Adams is pursuing her Ph.D. in Bioengineering at Rice University under the mentorship of Professor Athanasiou. Her thesis project focuses on using dermis-derived cells toward tissue engineering a spectrum of cartilages. She holds a B.S. in biomedical engineering and mechanical engineering from Duke University.

Printed in the United States
by Baker & Taylor Publisher Services